Pearson's

Pocket Guide to

Co

D0143599

Weber State University

Prentice Hall
Boston   Columbus   Indianapolis   New York   San Francisco
Upper Saddle River   Amsterdam   Cape Town   Dubai   London   Madrid   Milan
Munich   Paris   Montreal   Toronto   Delhi   Mexico City
Sao Paulo   Sydney   Hong Kong   Seoul   Singapore   Taipei   Tokyo

Editorial Director: Vernon R. Anthony
Acquisitions Editor: David Ploskonka
Editorial Assistant: Nancy Kesterson
Director of Marketing: David Gesell
Marketing Manager: Derril Trakalo
Senior Marketing Coordinator:
    Alicia Wozniak
Marketing Assistant: Les Roberts
Senior Project Manager: Maren L. Miller
Senior Managing Editor: JoEllen Gohr

Associate Managing Editor:
    Alexandrina Benedicto Wolf
Senior Operations Supervisor:
    Pat Tonneman
Operations Specialist: Deidra Skahill
Cover Designer: Suzanne Behnke
Cover Image: ©Anders Adermark/Fotolia
AV Project Manager: Janet Portisch
Composition: Kelly Barber
Printer/Binder: Edwards Brothers Malloy
Cover Printer: Edwards Brothers Malloy

Prentice Hall
is an imprint of

www.pearsonhighered.com

14 2019

ISBN 10: 0-13-215610-5
ISBN 13: 978-0-13-215610-3

Table of Contents

Chapter 1: Construction Math

Basic Trigonometry
Right Triangle

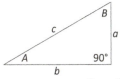

$$A + B = 90° \quad \blacklozenge \quad \sin A = \frac{\text{Opposite}}{\text{Hypotenuse}} = \frac{a}{c}$$

$$\cos A = \frac{\text{Adjacent}}{\text{Hypotenuse}} = \frac{b}{c} \quad \blacklozenge \quad \tan A = \frac{\text{Opposite}}{\text{Adjacent}} = \frac{a}{b}$$

$$c^2 = a^2 + b^2 \quad \text{Pythagorean Theorem}$$

Example: Determine angle B and the length of sides a and b for the following figure:

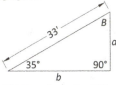

$$35° + B = 90° \text{ therefore } B = 90° - 35° = 55°$$

$$\sin 35° = \frac{a}{(33 \text{ ft})} \text{ therefore } a = (33 \text{ ft}) \sin 35° = 18.93 \text{ ft}$$

$$\cos 35° = \frac{b}{(33 \text{ ft})} \text{ therefore}$$

$$b = (33 \text{ ft}) \cos 35° = 27.03 \text{ ft}$$

Example: The two short legs of a right triangle are 12 feet and 9 feet long. Determine the length of its hypotenuse.

Use the Pythagorean Theorem:
$$c^2 = (12 \text{ ft})^2 + (9 \text{ ft})^2 = 225 \text{ ft}^2$$
$$c = \sqrt{225 \text{ ft}^2} = 15 \text{ ft}$$

Oblique Triangle

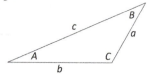

$$A + B + C = 180°$$
$$a^2 = b^2 + c^2 - 2bc \cos A \qquad \text{Law of Cosines}$$
$$\frac{a}{\sin A} = \frac{b}{\sin B} = \frac{c}{\sin C} \qquad \text{Law of Sines}$$

Example: Determine the length of side *a* for the following figure:

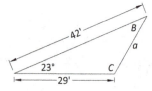

Use the Law of Cosines:
$$a^2 = (29 \text{ ft})^2 + (42 \text{ ft})^2 - 2(29 \text{ ft})(42 \text{ ft})\cos 23°$$
$$a = 19.04 \text{ ft}$$

2

Example: Determine angles *A* and *B* and the length of side *b* for the following figure:

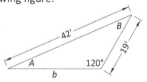

Use the Law of Sines to find angle *A*:

$$\frac{19 \text{ ft}}{\sin A} = \frac{42 \text{ ft}}{\sin 120°}$$

$$\sin A = \frac{(19 \text{ ft}) \sin 120°}{42 \text{ ft}} = 0.39177$$

$$A = 23.0649°$$

Find angle *B*:

$$23.0649° + B + 120° = 180°$$

$$B = 180° - 23.0649° - 120° = 36.9351°$$

Use the Law of Sines to find *b*:

$$\frac{b}{\sin 36.9351°} = \frac{42 \text{ ft}}{\sin 120°}$$

$$b = \frac{(42 \text{ ft}) \sin 36.9351°}{\sin 120°} = 29.14 \text{ ft}$$

Degree-Minutes-Seconds to Decimal Conversion

$$\text{Angle} = \text{Degrees} + \frac{\text{Minutes}}{60 \text{ min/deg}} + \frac{\text{Seconds}}{3,600 \text{ sec/deg}}$$

Example: Convert 32 degrees 22 minutes 15 seconds to degrees in decimal format.

$$\text{Angle} = 32 \text{ deg} + \frac{22 \text{ min}}{60 \text{ min/deg}} + \frac{15 \text{ sec}}{3,600 \text{ sec/deg}}$$

$$\text{Angle} = 32.3708°$$

Decimal to Degree-Minute-Second Conversion
Steps:
1. Record the numbers to the left of the decimal point as the number of degrees
2. Multiply the remainder by 60 min/deg
3. Record the numbers to the left of the decimal as the number of minutes
4. Multiply the remainder by 60 sec/min
5. Record the remainder as the number of seconds

Example: Convert 42.89562 degrees to degrees, minutes, and seconds.

Step 1:

42 degrees

Step 2:

$$0.89562 \text{ deg} \times 60 \text{ min/deg} = 53.7372 \text{ min}$$

Step 3:

53 minutes

Step 4:

$$0.7372 \text{ min} \times 60 \text{ sec/min} = 44.23 \text{ sec}$$

Step 5:

44.23 seconds
42 degrees 53 minutes 44.23 seconds

Properties of Common Two-Dimensional Shapes

Circle

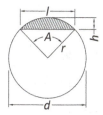

$$d = 2r \quad \blacklozenge \quad A \text{ is measured in degrees}$$

$$\text{Perimeter} = 2\pi r = \pi d \quad \blacklozenge \quad \text{Area} = \pi r^2 = \frac{\pi d^2}{4}$$

$$l = 2r\sin\frac{A}{2} = d\sin\frac{A}{2} \quad \blacklozenge \quad h = r\left(1 - \cos\frac{A}{2}\right) = \frac{l}{2}\tan\frac{A}{4}$$

$$\text{Arc length} = 2\pi r\frac{A}{360°} = \pi d\frac{A}{360°}$$

$$\text{Area}_{\text{Circular Segment}} = r^2\left[\frac{\pi A}{360°} - \cos\frac{A}{2}\sin\frac{A}{2}\right]$$

Example: Find the perimeter and area for a circle with a radius of 15 inches. Determine l, h, the arch length, and area of the circular segment (shaded area) for an angle of 120 degrees.

$$\text{Perimeter} = 2\pi(15 \text{ in}) = 94.25 \text{ in}$$

$$\text{Area} = \pi(15 \text{ in})^2 = 706.86 \text{ in}^2$$

$$l = 2(15 \text{ in})\sin\frac{120°}{2} = 25.98 \text{ in}$$

$$h = (15 \text{ in})\left(1 - \cos\frac{120°}{2}\right) = 7.50 \text{ in}$$

$$\text{Arc length} = 2\pi(15 \text{ in})\frac{120°}{360°} = 31.42 \text{ in}$$

$$\text{Area}_{\text{Circular Segment}} = (15 \text{ in})^2\left(\frac{\pi 120°}{360°} - \cos\frac{120°}{2}\sin\frac{120°}{2}\right)$$

$$\text{Area}_{\text{Circular Segment}} = 138.19 \text{ in}^2$$

Rectangle

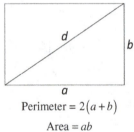

$$\text{Perimeter} = 2(a+b)$$
$$\text{Area} = ab$$
$$d = \sqrt{a^2 + b^2}$$

Example: Find the perimeter, area, and diagonal of a rectangle whose sides are 12 inches and 15 inches.

$$\text{Perimeter} = 2(12 \text{ in} + 15 \text{ in}) = 54 \text{ in}$$
$$\text{Area} = (12 \text{ in})(15 \text{ in}) = 180 \text{ in}$$
$$d = \sqrt{(12 \text{ in})^2 + (15 \text{ in})^2} = 19.21 \text{ in}^2$$

Square

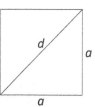

Perimeter $= 4a$

Area $= a^2$

$d = a\sqrt{2}$

Example: Find the perimeter, area, and diagonal of a square whose sides are 15 inches.

$$\text{Perimeter} = 4(15\text{ in}) = 60\text{ in}$$

$$\text{Area} = (15\text{ in})^2 = 225\text{ in}^2$$

$$d = (15\text{ in})\sqrt{2} = 21.21\text{ in}$$

Triangle

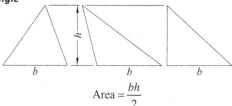

$$\text{Area} = \frac{bh}{2}$$

Example: Find the area of a triangle whose base is 15 inches and whose height is 12 inches.

$$\text{Area} = \frac{(15\text{ in})(12\text{ in})}{2} = 90\text{ in}^2$$

Trapezoid

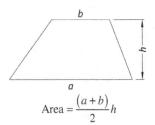

$$\text{Area} = \frac{(a+b)}{2}h$$

Example: Find the area of a trapezoid whose parallel sides are 15 and 22 inches and whose height is 12 inches.

$$\text{Area} = \frac{(15 \text{ in} + 22 \text{ in})}{2}(12 \text{ in}) = 222 \text{ in}^2$$

Parallelogram

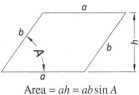

$$\text{Area} = ah = ab\sin A$$

Example: Find the area of a parallelogram whose base is 15 inches and whose height is 12 inches.

$$\text{Area} = (15 \text{ in})(12 \text{ in}) = 180 \text{ in}^2$$

Example: Find the area of a parallelogram whose sides are 12 and 15 inches. The sides intersect at a 45 degree angle.

$$\text{Area} = (15 \text{ in})(12 \text{ in})(\sin 45°) = 127.28 \text{ in}^2$$

Regular Polygons

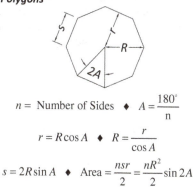

$n =$ Number of Sides $\quad \blacklozenge \quad A = \dfrac{180°}{n}$

$r = R\cos A \quad \blacklozenge \quad R = \dfrac{r}{\cos A}$

$s = 2R\sin A \quad \blacklozenge \quad \text{Area} = \dfrac{nsr}{2} = \dfrac{nR^2}{2}\sin 2A$

Example: Find the area of an octagon inscribed inside a circle with a radius of 15 inches. What are the lengths of its sides and the distance measured perpendicular from the sides to the center (r)?

$A = \dfrac{180°}{8} = 22.5°$

$s = 2(15 \text{ in})\sin 22.5° = 11.48 \text{ in}$

$r = (15 \text{ in})\cos 22.5° = 13.86 \text{ in}$

$\text{Area} = \dfrac{8(15 \text{ in})^2}{2}\sin(2 \times 22.5°) = 636.40 \text{ in}^2$

$\text{Area} = \dfrac{8(11.48 \text{ in})(13.86 \text{ in})}{2} = 636.45 \text{ in}^2$

Example: Find the area of a pentagon circumscribed outside a circle with a radius of 15 inches. What are the lengths of its sides and the distance measured from the center to the intersection of its sides (R)?

$$A = \frac{180°}{5} = 36° \quad \blacklozenge \quad R = \frac{15 \text{ in}}{\cos 36°} = 18.54 \text{ in}$$

$$s = 2(18.54 \text{ in})\sin 36° = 21.80 \text{ in}$$

$$\text{Area} = \frac{5(18.54 \text{ in})^2}{2}\sin(2 \times 36°) = 817.27 \text{ in}^2$$

$$\text{Area} = \frac{5(21.80 \text{ in})(15 \text{ in})}{2} = 817.50 \text{ in}^2$$

Ellipse

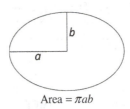

$$\text{Area} = \pi ab$$

Example: Find the area of an ellipse inscribed in a 12-inch by 15-inch rectangle.

$$\text{Area} = \pi\left(\frac{12 \text{ in}}{2}\right)\left(\frac{15 \text{ in}}{2}\right) = 141.37 \text{ in}^2$$

Square-Less-a-Quarter-Circle

A square-less-a-quarter-circle occur when corners are rounded, such as in a parking lot.

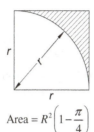

$$\text{Area} = R^2\left(1 - \frac{\pi}{4}\right)$$

Example: A quarter circle is inscribed inside a square whose sides are 12 inches. Find the area of the square less a quarter circle.

$$\text{Area} = (12 \text{ in})^2\left(1 - \frac{\pi}{4}\right) = 30.90 \text{ in}^2$$

Complex Shapes

Steps:

1. Divide the shape into triangles, squares, circles, etc.
2. Determine the area of each shape.
3. Add or subtract each shape to get the total area.

Example: Find the area of the shape shown below:

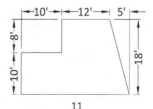

Step 1: Divide the figure as shown below:

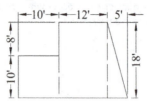

Step 2: Determine the area of each shape:

$$\text{Area}_{\text{Square}} = \left(10 \text{ in}\right)^2 = 100 \text{ in}^2$$

$$\text{Area}_{\text{Rectangle}} = \left(12 \text{ in}\right)\left(18 \text{ in}\right) = 216 \text{ in}^2$$

$$\text{Area}_{\text{Triangle}} = \frac{\left(5 \text{ in}\right)\left(18 \text{ in}\right)}{2} = 45 \text{ in}^2$$

Step 3: Add the shape together to get the total area:

$$\text{Area}_{\text{Total}} = 100 \text{ in}^2 + 216 \text{ in}^2 + 45 \text{ in}^2 = 361 \text{ in}^2$$

Irregular Polygons using the Coordinate Method
Steps:

1. Set up a table with four columns.
2. Proceeding around the polygon, list the X-coordinate in the second column and the Y-coordinate in the third column. The first point should be listed in the top and bottom rows of the table.
3. Multiply the second X-coordinate by the first Y-coordinate and place the resultant in the first column next to the second coordinate. Continue down the table until you reach the bottom of the table.

4. Multiply the first X-coordinate by the second Y-coordinate and place the resultant in the fourth column next to the second coordinate. Continue down the table until you reach the bottom of the table.
5. Sum the first and fourth columns.
6. The area is half the difference between the sum of the first and fourth columns.

Example: Find the area of the shape shown below. The coordinates are in inches:

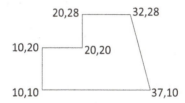

Steps 1 through 5 are shown in the following figure.

	X	y	
	10	10	
370 ← 37		10 →	100
320 ← 32		28 →	1,036
560 ← 20		28 →	896
560 ← 20		20 →	400
200 ← 10		20 →	400
200 ← 10		10 →	100
2,210			2,932

Step 6: The area is determined as follows:

$$\text{Area} = \frac{2,932 \text{ in}^2 - 2,210 \text{ in}^2}{2} = 361 \text{ in}^2$$

Volumes of Common Geometric Shapes
Spheres

Surface Area $= 4\pi r^2 = \pi d^2$ ◆ Volume $= \dfrac{4\pi r^3}{3} = \dfrac{\pi d^3}{3}$

Example: Find the surface area and the volume of a sphere with a radius of 15 inches.

$$\text{Surface Area} = 4\pi(15 \text{ in})^2 = 2,827 \text{ in}^2$$

$$\text{Volume} = \frac{4\pi(15 \text{ in})^3}{3} = 14,137 \text{ in}^3$$

Spherical Sector

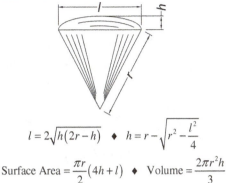

$$l = 2\sqrt{h(2r-h)} \quad \blacklozenge \quad h = r - \sqrt{r^2 - \frac{l^2}{4}}$$

$$\text{Surface Area} = \frac{\pi r}{2}(4h+l) \quad \blacklozenge \quad \text{Volume} = \frac{2\pi r^2 h}{3}$$

Example: Find the surface area, the volume, and *h* of a spherical sector with a radius of 15 inches and a diameter (*l*) of 18 inches on the surface of the sphere.

$$h = (15 \text{ in}) - \sqrt{(15 \text{ in})^2 - \frac{(18 \text{ in})^2}{4}} = 3 \text{ in}$$

$$\text{Surface Area} = \frac{\pi(15 \text{ in})}{2}(4\times3 \text{ in} + 18 \text{ in}) = 707 \text{ in}^2$$

$$\text{Volume} = \frac{2\pi(15 \text{ in})^2(3 \text{ in})}{3} = 1,414 \text{ in}^3$$

Spherical Segments

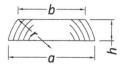

Surface area including top, bottom, and sides:

$$\text{Surface Area} = \frac{\pi}{4}\left(8rh + a^2 + b^2\right)$$

Area of sides only (excludes top and bottom):

$$\text{Surface Area} = \frac{\pi}{4}\left(8rh\right)$$

$$\text{Volume} = \frac{\pi h}{24}\left(3a^2 + 3b^2 + 4h^2\right)$$

Example: Find the surface area (including the top and bottom) and the volume of a spherical segment cut from a sphere with a radius of 15 inches. The sphere is cut at the midline of the sphere and 9 inches from the midline. The top of the spherical segment has a radius of 12 inches.

$a = 2r = 2 \times 15 \text{ in} = 30 \text{ in}$

$b = 2 \times 12 \text{ in} = 24 \text{ in}$

$$\text{Surface Area} = \frac{\pi}{4}\left(8(15 \text{ in})(9 \text{ in}) + (30 \text{ in})^2 + (24 \text{ in})^2\right)$$

$$\text{Surface Area} = 2,007 \text{ in}^2$$

$$\text{Volume} = \frac{\pi(9 \text{ in})}{24}\left(3(30 \text{ in})^2 + 3(24 \text{ in})^2 + 4(9 \text{ in})^2\right)$$

$$\text{Volume} = 5,598 \text{ in}^3$$

Right Cylinder

$$\text{Area}_{\text{Ends}} = 2\pi r^2 = \frac{\pi d^2}{2} \quad \blacklozenge \quad \text{Area}_{\text{Sides}} = 2\pi rh = \pi dh$$

$$\text{Volume} = \pi r^2 h$$

Example: Find the surface area and the volume of a right cylinder with a radius of 15 inches and a height of 24 inches.

$$\text{Area}_{\text{Ends}} = 2\pi(15 \text{ in})^2 = 1{,}414 \text{ in}^2$$

$$\text{Area}_{\text{Sides}} = 2\pi(15 \text{ in})(24 \text{ in}) = 2{,}262 \text{ in}^2$$

$$\text{Surface Area} = 1{,}414 \text{ in}^2 + 2{,}262 \text{ in}^2 = 3{,}676 \text{ in}^2$$

$$\text{Volume} = \pi(15 \text{ in})^2(24 \text{ in}) = 16{,}965 \text{ in}^3$$

Rectangular Column

$$\text{Area}_{\text{Ends}} = 2ab \quad \blacklozenge \quad \text{Area}_{\text{Sides}} = 2(a+b)h$$

$$\text{Volume} = abh$$

17

Example: Find the surface area and the volume of a rectangular column with a base of 12 inches by 15 inches and a height of 24 inches.

$$\text{Area}_{\text{Ends}} = 2(12 \text{ in})(15 \text{ in}) = 360 \text{ in}^2$$

$$\text{Area}_{\text{Sides}} = 2(12 \text{ in} + 15 \text{ in})(24 \text{ in}) = 1,296 \text{ in}^2$$

$$\text{Surface Area} = 360 \text{ in}^2 + 1,296 \text{ in}^2 = 1,656 \text{ in}^2$$

$$\text{Volume} = (12 \text{ in})(15 \text{ in})(24 \text{ in}) = 4,320 \text{ in}^3$$

Triangular Prism

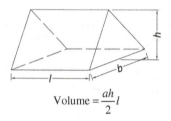

$$\text{Volume} = \frac{ah}{2}l$$

Example: Find the volume of a triangular prism. The ends of the prism have a base of 12 inches and a height of 15 inches. The length of the prism is 24 inches.

$$\text{Volume} = \frac{(12 \text{ in})(15 \text{ in})}{2}(24 \text{ in}) = 2,160 \text{ in}^3$$

Prism

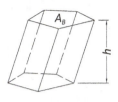

$$\text{Volume} = A_B h$$

Example: Find the volume of prism whose base is an octagon inscribed inside a circle with a radius of 15 inches and whose height, measured perpendicular to the base, is 48 inches.

The area of the octagon is found as follows:

$$A = \frac{180°}{n} = \frac{180°}{8} = 22.5°$$

$$\text{Area} = \frac{nR^2}{2}\sin 2A = \frac{8(15 \text{ in})^2}{2}\sin(2 \times 22.5°) = 636 \text{ in}^2$$

The volume of the prism is as follows:

$$\text{Volume} = (636 \text{ in}^2)(48 \text{ in}) = 30,528 \text{ in}^3$$

Circular Cone

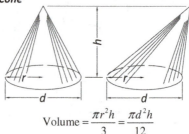

$$\text{Volume} = \frac{\pi r^2 h}{3} = \frac{\pi d^2 h}{12}$$

Example: Find the volume of a circular cone whose base has a diameter of 15 inches and whose height, measured perpendicular to the base, is 24 inches.

$$\text{Volume} = \frac{\pi (15 \text{ in})^2 (24 \text{ in})}{12} = 1,414 \text{ in}^3$$

Rectangular Pyramid

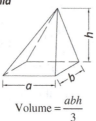

$$\text{Volume} = \frac{abh}{3}$$

Example: Find the volume of a rectangular pyramid with a base of 12 inches by 15 inches and a height of 24 inches, which is measured perpendicular to the base.

$$\text{Volume} = \frac{(12 \text{ in})(15 \text{ in})(24 \text{ in})}{3} = 1,440 \text{ in}^3$$

General Cone

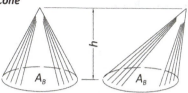

Note: The base of the cone may be any shape, not just round as shown here.

$$\text{Volume} = \frac{A_B h}{3}$$

Example: Find the volume of a cone with a triangular base. The triangular base of the cone has a base of 12 inches and a height of 15 inches. The height of the cone is 24 inches, which is measured perpendicular to the base.

The area of the base is as follows:

$$A_B = \frac{bh}{2} = \frac{(12 \text{ in})(15 \text{ in})}{2} = 90 \text{ in}^2$$

The volume is as follows:

$$\text{Volume} = \frac{(90 \text{ in}^2)(24 \text{ in})}{3} = 720 \text{ in}^3$$

Frustum of a General Cone

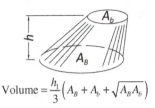

$$\text{Volume} = \frac{h_1}{3}\left(A_B + A_b + \sqrt{A_B A_b}\right)$$

Example: Find the volume of a frustum of a cone whose bottom surface has a radius of 15 inches and whose top surface has a radius of 12 inches. Its height is 24 inches and is measured perpendicular to the ends.

The areas of the bases are as follows:

$$A_B = \pi r_B^2 = \pi (15 \text{ in})^2 = 707 \text{ in}^2$$
$$A_b = \pi r_b^2 = \pi (12 \text{ in})^2 = 452 \text{ in}^2$$

The volume is as follows:

$$\text{Volume} = \frac{(24 \text{ in})}{3} \times$$
$$\left((707 \text{ in}^2) + (452 \text{ in}^2) + \sqrt{(707 \text{ in}^2)(452 \text{ in}^2)}\right)$$
$$\text{Volume} = 13,794 \text{ in}^3$$

Ellipsoid

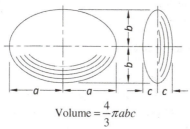

$$\text{Volume} = \frac{4}{3}\pi abc$$

Example: Find the volume of an ellipsoid inscribed in a 12-inch by 15-inch by 24-inch rectangular column.

$$a = \frac{12 \text{ in}}{2} = 6 \text{ in}$$

$$b = \frac{15 \text{ in}}{2} = 7.5 \text{ in}$$

$$c = \frac{24 \text{ in}}{2} = 12 \text{ in}$$

$$\text{Volume} = \frac{4}{3}\pi (6 \text{ in})(7.5 \text{ in})(12 \text{ in}) = 2,262 \text{ in}^3$$

Conversion factors

Length (English)

1 foot = 12 inches
1 yard = 3 feet = 36 inches
1 rod = 5.5 yards
1 furlong = 40 rods = 220 yards = 660 feet
1 mile = 1,760 yards = 5,280 feet

Foot Equivalents of Inches and Fractions of an Inch

Fraction	Decimal	Fraction	Decimal
1/32	0.031	17/32	0.531
1/16	0.063	9/16	0.563
3/32	0.094	19/32	0.594
1/8	0.125	5/8	0.625
5/32	0.156	21/32	0.656
3/16	0.188	11/16	0.688
7/32	0.219	23/32	0.719
1/4	0.250	3/4	0.750
9/32	0.281	25/32	0.781
5/16	0.313	13/16	0.813
11/32	0.344	27/32	0.844
3/8	0.375	5/8	0.625
13/32	0.406	29/32	0.906
7/16	0.438	15/16	0.938
15/32	0.469	31/32	0.969
1/2	0.500		

Area - Acres, Square Yards, Square Feet, and Square Inches

1 foot^2 = 144 inches^2
1 yard^2 = 9 feet^2 = 1,296 inches^2
1 rod^2 = 30 1/4 yards^2
1 acre = 160 rods^2 = 4,840 yard^2 = 43,560 feet^2
1 mile^2 = 640 acres = 3,097,600 yards^2

Volume (English)

1 yard^3 = 27 feet^3
1 foot^3 = 1,728 inches^3
1 gallon (US liquid) = 4 quarts = 8 pints = 128 fluid ounces
1 gallon (US liquid) = 231 inches^3
1 foot^3 = 7.48 gallons (US)

Weight (English)

1 pound = 16 ounces = 256 dram = 7,000 grains
1 short hundredweight = 100 pounds
1 long hundredweight = 112 pounds
1 short ton = 2,000 pounds
1 long ton = 2,240 pounds

Metric to English Conversions

1 centimeter = 0.39 inch
1 meter = 1.094 yards = 3.28 feet = 39.37 inches
1 kilometer = 0.62 mile
1 centimeter^2 = 0.155 inch^2
1 hectare = 2.47 acres
1 kilometer^2 = 0.386 mile^2
1 centimeter^3 = 0.06 inch^3
1 meter^3 = 1.31 yards^3
1 milliliter = 0.034 fluid ounce (US)
1 liter = 1.06 quarts (US)
1 milligram = 0.015 grain
1 kilogram = 2.205 pounds
1 metric ton = 1.1 short tons = 2,204 pounds

English to Metric

1 inch = 0.0254 meters = 2.54 centimeters
1 foot = 0.305 meter
1 yard = 0.914 meter
1 mile = 1.609 kilometers
1 inch^2 = 6.45 centimeters^2
1 mile^2 = 2.59 kilometers^2
1 acre = 0.405 hectare
1 inch^3 = 16.4 centimeters^3
1 yard^3 = 0.765 meter^3
1 fluid ounce (US) = 29.6 milliliters
1 quart (US) = 0.946 liter
1 gallon (US) = 3.8 liters
1 grain = 64.8 milligrams
1 pound = 0.45 kilogram
1 short ton = 0.91 metric ton

Mechanical and Electrical Conversion Factors

1 ton AC = 12,000 BTU
1 watt = 3.412 BTU/hr
1 BTU/hr = 0.2931 watts
1 gallon/minute = 0.06309 liters/second
1 foot^3/minute = 0.4719 liters/second
1 liters/second = 15.85 gallon/minute = 2.119 foot^3/minute
1 foot^3 of water (@32°F) = 62.416 pounds
1 foot^3 of water (@100°F) = 61.998 pounds
1 foot^3 of water (customary) = 62.4 pounds
1 gallon of water (customary) = 8.33 pounds

Board Feet

A board foot is the amount of wood contained in a 12-inch by 12-inch by 1-inch board. It is calculated using nominal dimensions with the following equation:

$$\text{Board foot} = \frac{(\text{Thickness}_{\text{Inches}})(\text{Width}_{\text{Inches}})(\text{Length}_{\text{Feet}})}{12}$$

Example: Determine the number of board feet (bft) in a 10-foot-long 2x8.

$$\text{Board foot} = \frac{(2)(8)(10)}{12} = 13.33 \text{ bft}$$

Pitch and Slope

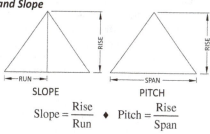

SLOPE   PITCH

$$\text{Slope} = \frac{\text{Rise}}{\text{Run}} \quad \blacklozenge \quad \text{Pitch} = \frac{\text{Rise}}{\text{Span}}$$

Slope	Pitch	Slope	Pitch
1:12	1/24	7:12	7/24
2:12	1/12	8:12	1/3
3:12	1/8	9:12	9/24
4:12	1/6	10:12	5/12
5:12	5/24	11:12	11/24
6:12	1/4	12:12	1/2

Chapter 2: The Project Manual

The project manual is a written document that accompanies the plans or drawings. It is often mistakenly referred to as the specifications, because it contains the technical specifications, but it contains much more. The project manual sets many of the requirements that the contractor must meet in order for the project to be a success.

Contents of the Project Manual

A typical project manual for a design-bid-build project contains the following:

The **Invitation to Bid** gives a brief description of the project and invites contractors to bid on the project. It includes the contact information for the architect or engineer, the bid date, restrictions on bidders, expected price range, and the expected duration of the project. The invitation to bid is often posted or published in the newspaper for public projects.

The **Instructions to Bidders** are instructions that the contractor must follow to submit a complete bid. Failure to follow these instructions may result in the bid begin disqualified. The instructions to bidder may require the bidder to attend a pre-bid meeting in order to be qualified to bid.

The **Bid Documents** includes all of the forms the contactor must submit for the bid and may include bid forms, a schedule of values, and contractor certifications (e.g., certification that they are a qualified minority business).

The **Bonds** include the bonding requirements for the project and often include the bond forms that must be used for the project. Bonds are discussed in detail later in this chapter.

The **Contract** is the agreement between the owner (or the owner's representative) and the contractor for the construction of the project. If the contractor is the successful bidder, she will need to sign this contract or risk losing the bid bond. The types of contracts are discussed later in this chapter.

The **General Conditions** establish conditions that apply to the entire project. The general conditions address such issues as project supervision, warranty, taxes, permits and fees, allowances, project schedule, shop drawings and samples, cleanup, contract administrative procedures, claims and dispute procedures, subcontracting of work, work by the owner, change orders, time of completion, time extensions, payment procedures, required certification, substantial completion, safety, bonds, insurance, correction of work, testing and inspections, and termination of the contract. The general conditions are often a standard document, such as American Institute of Architects (AIA) form A201. Insurance is discussed later in this chapter.

The **Supplementary or Special Conditions** are used to modify the general conditions, by deleting provisions from and adding provisions to the general conditions to meet the project's specific requirements.

The **Technical Specifications** or specifications define the quality of the materials used on the project and their performance. The specifications are often organized by the Construction Specifications Institute's (CSI's) MasterFormat™. The MasterFormat™ is show later in this chapter. The specifications are divided into three parts.

Part 1 – General identifies the scope of work for the specifications and includes a description of the work; a list of related work found in other specifications; the quality assurance requirements; the submittals required; and product delivery, storage, and handling.

Part 2 – Products identifies the products that are acceptable. This may be done by setting a standard for the product or calling out approved products from specific manufacturers.

Part 3 – Execution covers the installation requirements for the product. This may include step-by-step instructions or simply require that the manufacturer's instructions be followed.

Other items (e.g., soils report) may be included in the project manual, while other items (e.g., testing standards) are often included by referencing the document, rather than including them in the project manual. Documents that are included by reference are treated as if they are attached to the document even though they are not.

Addendums modify the project manual and drawings before the bid, and must be acknowledged on the bid.

Contract Types

Construction contracts may be divided into four broad categories (design bid build, design build, integrated project delivery, and construction manager) based on the relationships between the parties. The relationship between two parties may be contractual, where there is a signed contract between the parties that identifies their roles and obligations; or the relationship may be information, where the parties share information and work together based upon a contractual relationship that each of the parties has with a third party (such as the project's owner). The four broad categories are discussed below, with the contractual relationships shown as double lines and information relationships shown as single lines.

The **Design-Bid-Build** contract is the most common type of contract. Under this type of contract, the owner hires an architect or engineer (the design consultant) to design the project. Once the design is complete, the project is bid and the contractor is hired by the owner to construct the project. During construction, the architect or engineer administers the contract on behalf of the owner. The typical relationships in a design-bid-build contract are show in the following figure.

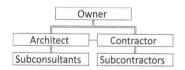

Using a **Design-Build** contract, the owner hires a single entity to both design and build the project. The advantage of this is that the contractor's experience and knowledge of costs and construction methods can be incorporated into the design. The design-build entity may be a single company that employs both designers and construction personnel or it may be a joint venture between companies. The typical relationships in a design-build contract are shown in the following figure.

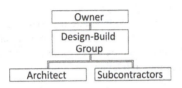

With an **Integrated Project Delivery** (IPD) contract, the owner hires the design professionals, the construction manager (often the contractor), and key subcontractors early in the design process. Like a Design-Build, IPD has the advantage of incorporating the contractor's and subcontractor's knowledge of costs and construction methods into the design. Unlike Design-Build, the owner retains greater control over the design. The typical relationships in an IPD contract are shown in the following figure.

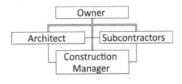

Using a **Construction Manager** (CM) contract, the owner hires a construction manager to act as its agent for the design and construction of the project. The CM is hired early in the design process. There are two types of construction manager contracts.

Under an **Agency CM** contract, the owner retains the risk of cost overruns and the construction manager simply acts as an agent for the owner, coordinating the work of the architect and the contractor. Under an Agency CM contract the construction manager may hire the subcontractors on behalf of the owner, or the subcontractors may be hired by the contractor, or both. The typical relationships in an agency CM contract are shown in the following figure.

Under a **CM-at-Risk** contract, the construction manager acts as the contractor and assumes the risk of cost overruns, which may be in the form of a guaranteed maximum price (GMP) to the owner. The typical relationships in a CM-at-risk contract are shown in the following figure.

Contract Compensation

There are a number of different ways to establish the contractor compensation for the construction project.

Under a **lump-sum**, **stipulated-sum**, or **fixed-price** contract, the contractor is paid a set price which is established through competitive bidding or negotiations. If the scope of the project changes, a change order is issued that changes the project's price and time of completion.

Under a **Cost-Plus** contract, the contractor is paid the construction costs plus a fee for managing the project. The fee may be a percentage of the costs, a flat fee, a sliding scale where the percentage decreases as the costs increase, or a fixed fee with penalties for costs increasing and a bonus for costs decreasing. How the fee is set is only limited by one's imagination.

A **Guaranteed-Maximum-Price** (G-Max or GMP) contract is a cost-plus contract where the contractor guarantees the owner that the contraction costs will not exceed a specified price. Should the cost exceed this price, the contractor will pay for the cost overrun out of his pocket.

Unit-Price contracts are commonly used in road and utility construction. Under a unit-price contract, the contractor is paid a set price for each unit of work (e.g., cubic yard of dirt excavated, foot of pipe installed). The amount the contractor is paid is calculated by determining the actual quantity of work performed and multiplying it by the unit price.

Sources of Contracts

There are two common sources of standardized construction contracts. They are as follows:

American Institute of Architects (AIA)

For more than 120 years the American Institute of Architects has been a primary provider of standard contracts to the architecture and construction industries. The AIA has over 100 standard contracts and forms. Some of the more common AIA forms are listed below:

No.	Use
A101	Stipulate sum contract
A102	Cost plus contract with GMP
A103	Cost plus contract without GMP
A201	General conditions
B101	Agreement between owner and architect
G701	Change order
G702	Payment application
G704	Substantial completion

For more information visit www.aia.org.

ConsensusDocs

ConsensusDocs has over 90 standard construction documents that have been indorsed by 23 construction associations including Associated General Contractors of America (AGC) and Associated Builders and Contractors, Inc. (ABC).

Visit consensusdocs.org for more information.

Bonds

Bonds are issued by a surety and guarantee that the surety will step in and fulfill the contractor's contractual obligations should the contractor fail to do so. Unlike insurance policies, the surety will seek to recover losses from the contractor. There are three types of bonds commonly used on a construction project.

The **Bid Bond** guarantees that if a contractor is awarded the bid within the time specified, the contractor will enter into the contract and provide the required bonds. Should the contractor fail to do so, the bid bond is forfeited. Typically the amount of the bid bond is 5% to 10% of the bid.

The **Payment Bond** guarantees that the contractor will pay the labor, subcontractors, and material suppliers on the project. In the event that the contractor does not pay these people, the surety will step in and make the payments.

The **Performance Bond** guarantees that the contractor will complete the construction project in accordance with the contract. Should the contractor fail to do so, the surety will step in and fulfill the requirements of the contract.

The amount of the payment and performance bonds is equal to the contract amount. The cost of these bonds is typically 0.5% to 2% of the contract amount. The percentage varies based on the size of the project. The bid bond is often issued free of charge.

Visit www.sio.org for more information on bonds.

Insurance

Contractors must carry insurance to meet requirements of their construction contracts and to protect themselves against losses. The following is a list of the different types of insurance available to contractors.

Commercial General Liability insurance covers liability arising out of actions by the contractor and the company's employees, such as bodily injury and property damage or loss.

Automotive insurance covers vehicles for use on public roads including cars, trucks, portable office trailers used on the job site, and construction equipment driving on public roads, such as dump trucks.

Inland Marine insurance covers off-road construction equipment, such as backhoes, scrapers, and dump trucks not licensed for use on public roads.

Property insurance covers real property (real estate) owned by the contractor, such as a main office building.

Business Personal Property insurance covers the contents of a building, such as computer and furniture.

All of the above insurance is usually included in one policy.

Errors and Omissions insurance covers liability arising from errors or omissions by the designers of a project. Design-build contractors should carry error and omission insurance.

An **Umbrella** insurance policy goes on top of all the company's insurance, increasing the limits of coverage.

Workers' Compensation insurance protects the contractor against liability arising from injury to or death of employees. The contractor must maintain the minimum insurance required by state and federal law. The contract with the owner may require additional insurance coverage. See Chapter 6 for more information on workers' compensation insurance.

Builder's Risk insurance covers the construction project and may be purchased by the project's owner or the contractor. The contractor may be responsible for paying any deductibles even if they didn't purchase the insurance.

The construction contract sets the minimum amount for each insurable occurrence and an aggregate limit for all occurrences. It requires that the insurance indemnify (cover) the owner, architect, and engineer against liability arising out of actions of the contractor. This is done by naming the owner and others as an additionally insured on the insurance policy.

The contractor's insurance covers liability caused by subcontractors that is not covered by the subcontractor's insurance. Contractors should be named as an additionally insured on their subcontractor's insurance policies and verify that all their subcontractors carry the required insurance.

Divisions of the MasterFormat™ 2010 Edition

Division 00: Procurement and Contracting Requirements
Division 01: General Requirements
Division 02: Existing Conditions
Division 03: Concrete
Division 04: Masonry
Division 05: Metals
Division 06: Wood, Plastics, and Composites
Division 07: Thermal and Moisture Protection
Division 08: Openings
Division 09: Finishes
Division 10: Specialties
Division 11: Equipment
Division 12: Furnishings
Division 13: Special Construction
Division 14: Conveying Equipment
Division 21: Fire Suppression
Division 22: Plumbing
Division 23: Heating, Ventilating, and Air Conditioning (HVAC)
Division 25: Integrated Automation
Division 26: Electrical
Division 27: Communications
Division 28: Electronic Safety and Security
Division 31: Earthwork
Division 32: Exterior Improvements
Division 33: Utilities
Division 34: Transportation
Division 35: Waterway and Marine Construction
Division 40: Process Integration
Division 41: Material Processing and Handling Equipment

Division 42: Process Heating, Cooling, and Drying
Equipment
Division 43: Process Gas and Liquid Handling, Purification,
and Storage Equipment
Division 44: Pollution and Waste Control Equipment
Division 45: Industry-Specific Manufacturing Equipment
Division 46: Water and Wastewater Equipment
Division 48: Electrical Power Generation
Sections not listed above are reserved for future use.

The Numbers and Titles used in this textbook are from
MasterFormat™ 2010, published by The Construction
Specification Institute (CSI) and Construction Specifications
Canada (CSC), and are used with permission from CSI. For
those interested in a more in-depth explanation of
MasterFormat™ 2010 and its use in the construction
industry visit www.csinet.org/masterformat or contact:

The Construction Specifications Institute (CSI)
110 South Union Street, Suite 100
Alexandria, VA 22314
800-689-2900: 703-684-0300
www.csinet.org

Chapter 3: The Drawings

The drawings are graphical representations of the project, using lines and symbols to represent the various components. An understanding of the language of drawings (i.e., the lines and symbols) is necessary to read and understand the construction drawings.

Views

The three most common views used in construction drawings are orthographic, isometric, and oblique.

Orthographic drawings are prepared by projecting a view of the building onto a flat surface. Care must be used when dealing with surfaces that are not parallel to the drawing surface of an orthographic projection (e.g., a roof) because the actual lengths of the surface will be longer than shown on the drawings.

In **Isometric** drawings, vertical lines are drawn vertically on the page; the X-axis of the building is drawn at an angle of $30°$ to horizontal, moving up towards the right; the Y-axis of the building is drawn at an angle of $30°$ to horizontal, moving up towards the left. Isometric drawings are used to show objects in three dimensions.

In **Oblique** drawings, one surface of the object is drawn flat to the drawing's surface and parallel lines are included to add depth to the objects. Oblique drawings are often used to show the section of an object.

Line Types and Uses

The way a line is drawn changes the meaning of the line. The following is a list of the most common types of lines and their uses.

Object lines are used to show the parts of the building and are drawn with heavy solid lines.

_____

OBJECT LINE

Hidden lines are used to show the parts of the building that are hidden from view, but are necessary to construct the building (e.g., buried footings) or to show objects that are above the drawing surface (e.g., objects on the ceiling). Hidden lines are drawn as heavy dashed lines.

— — — — — — — — — — — — — — — — —

HIDDEN LINE

Centerlines mark the center of an object and are drawn with alternating short and long dashes. When an object is circular, the center is marked with two centerlines drawn at right angles to each other.

_____ — _____ — _____ — _____

CENTER LINE

Leader lines connect notes with objects on the drawings, showing where the note applies. Leader lines are drawn with a light solid line and often end with an arrowhead.

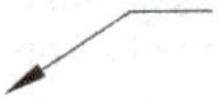

LEADER LINE

Dimension and **Extension** lines are used to show the length of a building component and are drawn with light solid lines. The dimension lines show the length of the object and end with arrowheads, slashes, or dots. The extension lines extend from just outside the object to past the end of the dimension line and show the limits of the object being measured.

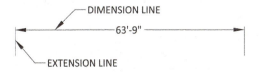

Cutting Planes show where a section passes through an object and are drawn with two short dashes and a long dash. Arrows, perpendicular to the cutting plane, are often drawn at the end of the cutting plane to show the direction of the view.

Break Lines show where building components have been left off of the drawings to make the drawings clearer (e.g., sheathing may be left off a wall to show its interior) or to reduce the size of a drawing (e.g., a center portion of a beam may be excluded, leaving the ends).

Scale

Drawings are drawn to scale. There are two types of scales used: the engineering and the architectural scale.

Engineering Scale

The engineering scale is used for site work, which is measured in feet and fraction of a foot. An engineering scale often includes the following scales, which can be multiplied or divided by 10 to get additional scales (e.g., 1"=100'):

1" = 10'	1" = 20'
1" = 30'	1" = 40'
1" = 50'	1" = 60'

An engineering scale is read just as you would read a ruler, by lining the zero of the scale (the left side of the scale) up with one end of the object and reading the length of the object off of the right side of the scale.

Example: What are the lengths of Lines A, B, and C in the below figure?

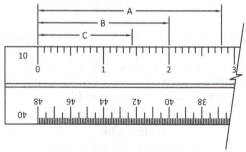

A = 28 feet B = 20 feet C = 14.3± feet

Architectural Scale

The architectural scale is used for building structures, which are measured in feet, inches, and fractions of an inch. An architectural scale often includes a standard ruler and the following scales:

3/32" = 1'	3/16" = 1'
1/8" = 1'	1/4" = 1'
3/8" = 1'	3/4" = 1'
1/2" = 1'	1" = 1'
1 1/2" = 1'	3" = 1'

Each side of an architectural scale contains two scales; one running left to right and one running right to left. One scale is twice the other scale. An architectural scale is read by lining one end of the object up with a foot mark with the other end so that it falls within the inch marks. The feet are read off of one end and the inches off of the other.

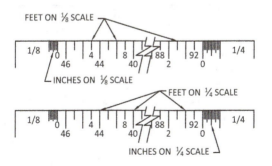

Example: What are the lengths of Lines A, B, C, and D in the below figure?

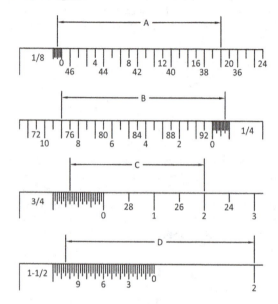

A = 19 feet 6 inches
B = 9 feet 9 inches
C = 2 feet 8 inches
D = 1 feet 10 1/2 inches

Drawing Types

The four most common types of drawings include plan views, elevations, sections, and details.

The **Plan View** shows the building or site from above and shows the location of objects in the horizontal (X-Y plane). For buildings, the plan view is typically drawn at an elevation of four feet above the floor. Objects that are hidden below another object and objects above the viewing elevation are drawn with hidden lines. Plan views include site, footing and foundation, floor, reflected ceiling, and structural framing plans.

Elevations show the heights or elevations of objects (Z axis). When combined with the plan view, the object can be located in all three dimensions (X, Y, and Z). Elevations are often shown for the exterior walls and interior walls where there is extensive finish work (e.g., trim, cabinets).

Sections show the building as if it had been cut along a cutting plane, showing only the objects that the cutting plane passes through. Objects in the background are not included in sections. The cutting plane is shown on the plan view or elevation to show where the section is located in relationship to the rest of the drawings, with arrows showing the direction the section is being viewed. Sections include building sections and wall sections.

Details show additional information about part of the project and are drawn at a larger scale than is used for the plan views and sections. The location of details is marked on the other drawing to show where the detail applies.

Drawings and Their Organization

The drawings for a construction project may contain the following drawings. The drawings are listed in the order they most commonly appear in the drawing package.

General Information sheets include the title page; sheets containing abbreviations, general notes, code information (e.g., occupancy and the occupancy load for each room), and survey information; and signature sheets used for approval of the drawings by the design professionals and code officials.

Site Drawings show the construction of all site features and the excavation required for the building. The site drawings include plan views and details. The site plan is usually drawn using the largest engineering scale that will fit on the page.

Architectural drawings show the design of the building. They include floor plans, elevations, building sections, reflected ceiling plans, wall sections, and details. The floor plans, elevations, and building sections are typically drawn at 1/4" = 1' or 1/8" = 1'.

Equipment drawings show the layout of special equipment (e.g., coolers in a store, kitchen equipment in a restaurant). Equipment drawings include plan views, elevations, sections, and details.

Structural drawings show how the building's structure is to be constructed. They show the design of the footings, foundation, and the building's supporting structure. Structural drawings include structural notes and abbreviations, plan views, sections, and details.

Mechanical drawings include the drawings for the heating, ventilation, and air conditioning (HVAC) system, the fire sprinkler system, and the plumbing. They show the design of these three systems. The mechanical drawings include mechanical notes, plan views, isometric line drawings, and details.

Electrical drawings show the design of the electrical system and include electrical notes, plan views, power diagrams, panel layouts with load calculations, electrical details, and lighting plans.

Drawing Log

A log or list of the most current drawings and their revision numbers must be kept at the jobsite. When a revised drawing is received, it should be recorded in the log and copies of the drawing should replace the old drawing in all of the plan sets. When a part of a drawing sheet is replaced, the replaced portion of the drawings should be marked out and a reference to the new drawings should be noted. This change should also be recorded in the drawing log.

Standard Symbols

Drawing symbols vary from architect to architect; however, they are often similar enough to be easily recognized. Below are some common symbols and how they may look.

Materials in Elevation

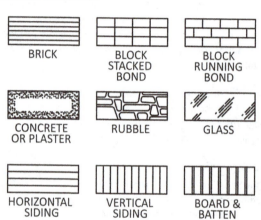

BRICK

BLOCK STACKED BOND

BLOCK RUNNING BOND

CONCRETE OR PLASTER

RUBBLE

GLASS

HORIZONTAL SIDING

VERTICAL SIDING

BOARD & BATTEN

Materials in Section

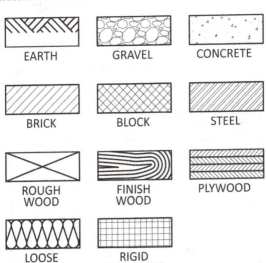

EARTH	GRAVEL	CONCRETE
BRICK	BLOCK	STEEL
ROUGH WOOD	FINISH WOOD	PLYWOOD
LOOSE OR BATT INSULATION	RIGID INSULATION	

Wielding Symbols

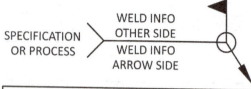

SPECIFICATION OR PROCESS

WELD INFO OTHER SIDE

WELD INFO ARROW SIDE

WELD ALL AROUND		
◢	FIELD WELD	
○	WELD ALL AROUND	
⌒	BACK	
◺	FILLET	
▢	PLUG/SLOT	
||	SQUARE	
∨	V	
⟋	BEVEL	
Y	U	
Ρ	J	
⋎	FLAIR V	
⌐	FLAIR BEVEL	

Doors and Windows

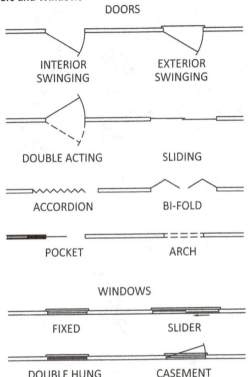

DOORS

INTERIOR SWINGING

EXTERIOR SWINGING

DOUBLE ACTING

SLIDING

ACCORDION

BI-FOLD

POCKET

ARCH

WINDOWS

FIXED

SLIDER

DOUBLE HUNG

CASEMENT

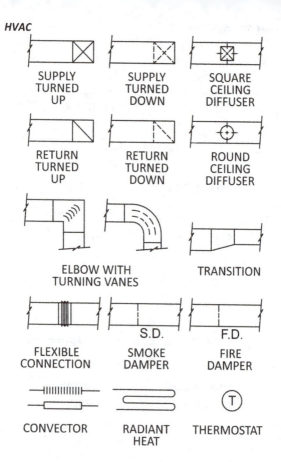

SUPPLY TURNED UP

SUPPLY TURNED DOWN

SQUARE CEILING DIFFUSER

RETURN TURNED UP

RETURN TURNED DOWN

ROUND CEILING DIFFUSER

ELBOW WITH TURNING VANES

TRANSITION

FLEXIBLE CONNECTION

S.D.

SMOKE DAMPER

F.D.

FIRE DAMPER

CONVECTOR

RADIANT HEAT

THERMOSTAT

Plumbing

FIXTURES

TUB   WHIRLPOOL   SHOWER

VANITY   WALL SINK   KITCHEN SINK

DW

DF

WH

DISH-WASHER   DRINKING FOUNTAIN   WATER HEATER

FLUSH WATER CLOSET   TANK WATER CLOSET   URINAL

PIPING

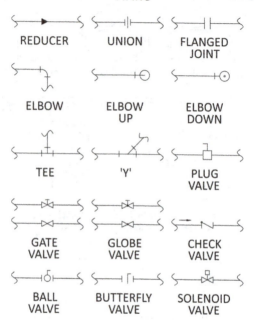

REDUCER

UNION

FLANGED
JOINT

ELBOW

ELBOW
UP

ELBOW
DOWN

TEE

'Y'

PLUG
VALVE

GATE
VALVE

GLOBE
VALVE

CHECK
VALVE

BALL
VALVE

BUTTERFLY
VALVE

SOLENOID
VALVE

Electrical

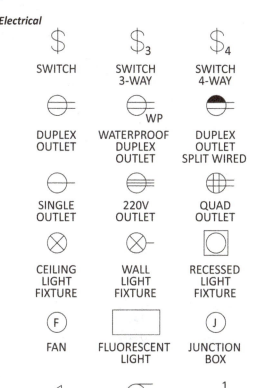

$	$₃	$₄
SWITCH	SWITCH 3-WAY	SWITCH 4-WAY
DUPLEX OUTLET	WATERPROOF DUPLEX OUTLET	DUPLEX OUTLET SPLIT WIRED
SINGLE OUTLET	220V OUTLET	QUAD OUTLET
CEILING LIGHT FIXTURE	WALL LIGHT FIXTURE	RECESSED LIGHT FIXTURE
FAN	FLUORESCENT LIGHT	JUNCTION BOX
TELEPHONE	MOTOR	HOME RUN TO PANEL

Paper sizes

The most common paper sizes for drawings are as follows:

Size	ANSI	Architectural
A	8.5″ × 11″	9″ × 12″
B	11″ × 17″	12″ × 18″
C	17″ × 22″	18″ × 24″
D	22″ × 34″	24″ × 36″
E	34″ × 44″	36″ × 48″

Roof types

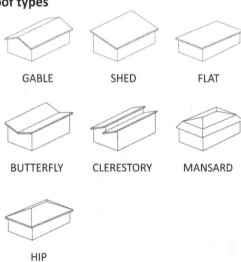

GABLE   SHED   FLAT

BUTTERFLY   CLERESTORY   MANSARD

HIP

Chapter 4: Construction Communications

The number one complaint bosses make about construction managers is that they are poor communicators. Poor communication is the source of many construction delays and cost increases. Poor communication leaves the client with the impression that the company is unprofessional. Strong communication skills are a must for any construction manager. The following is an overview of some of the more important forms of communication.

Phone Calls

A lot of a construction manager's time is spent on the phone coordinating construction activities. Before making a call, the caller should have a clear objective for the call (e.g., get the status of a submittal). The caller should speak clearly, be brief, and get to the point quickly. Phone calls should be returned promptly.

Leaving a Message

When leaving a message, the caller should identify herself, the company she works for, the project she is calling about, a brief message, date and time of the call, urgency, and the phone number where she can be reached. She should speak clearly and say her phone number slowly so that the listener can write it down without having to listen to the message again.

Phone Log

A written log of all phone calls should be kept. The log should include the date and time of the call, the participants, the topics discussed, and the promises made.

Letter Writing

All letters should leave the recipient with a good impression of the company and the sender. This is done by preparing a clearly written letter that is free of spelling, punctuation, and typographical errors. All letters should be proofread by another person. The following is an overview of the parts of a standard business letter in the order they appear in the letter and a sample letter.

Sender's Address includes the company's name and mailing address of the sender. This is omitted if the letter is printed on the company's letterhead.

The **Date** the letter was finished. The date should be written out.

Inside Address includes the name, company, and address of the recipient. Letters should be addressed to a specific person, not the company. The recipient's name should be spelled correctly and include the recipient's preferred title (e.g., Mr., Ms., Dr.) and credentials (e.g., PE, Ph.D.). If you are unsure of the person's preferred title and credentials, call the person's secretary to obtain the information.

The **Subject Line** should clearly identify the subject of the letter, so that it is easy to locate the letter in the future. The subject should include the project name and main topic(s) of the letter. The subject line often begins with "RE:", short for regarding.

The **Salutation** is a greeting to the recipient, such as "Dear Mr. Jones:"

The **Body** of the letter should be written in a clear, concise, and orderly manner. The letter should be organized with main points first, followed by lesser or supporting points. The end of the letter may contain a summary of the key points or request that the recipient perform some action (e.g., return the submittals by a specified date). The concluding paragraph should contain the name and contact information for the person that the receipt should contact if they have questions, which may be someone other than the person signing the letter.

The **Closing** consists of a closing statement (e.g., Sincerely, Best Regards) and the name and title of the person signing the letter.

If **Enclosures** are included with the letter, the word "Enclosure" or "Enclosures" appears after the closing. A list of the enclosures may be included.

CC: is followed by a list of the persons who are to receive a copy of the letter. The recipient is not included in the list.

Writer's and Typist's Initials are included at the bottom of the letter. The writer's (dictator's) are capitalized followed by the typist's initials which are written in lower case.

The **File Name and Location** of the electronic copy of the letter, so it may be found quickly.

For letters longer than two pages, the header of the second page should include the name of the recipient, the date of the letter, and the page number.

1802 University Circle
Ogden, Utah 84408-1802
May 27, 2010

Mr. John Robertson
Mt. Ogden Development
485 South Deer Born Lane
South Ogden, Utah

RE: Door Submittal for 10th Street Office Complex

Dear Mr. Robertson:

Enclosed are three copies of the catalogue cuts for the proposed doors and one set of veneer samples. Please review the submittals and let me know if they are acceptable. The submittals need to be approved by June 10 to maintain the agreed upon schedule.

If you have any questions, please contact me at 801-626-0000.

Sincerely,

Steven Peterson, MBA, PE

Enclosure

CC: file

Sample Letter – Block Format

E-mail Writing

E-mail is widely used in the construction industry, and is replacing letters in many cases. It is just as important for e-mails to be well written as it is for letters. E-mails should abide by the following rules:

- They should be clearly written, free of spelling, punctuation, and typographical errors. Spelling and grammar check should be used even if it means writing the e-mail in a word processor and copying it to the e-mail software. E-mails should not be written using all capital letters.

- They should only be sent to parties that need the information. Sending e-mails to large groups to cover your bases wastes their time and leaves a bad impression. Junk e-mails should not be forwarded.

- They should contain a subject line so that they can be quickly located. When the topic changes the subject line should change. E-mails with poor subject lines are often placed in the trash without even being opened.

- They should be brief and to the point. They should clearly state the desired action the sender wants the recipient to take.

- Emotion icons and text abbreviation should not be used.

- The signature block should contain the contact information for the sender, such as his phone number(s) and fax number.

Request for Information (RFI)

During the course of a project, the construction manager may need addition information from the architect, subcontractors, and suppliers. When the request is more complex than can be handled in a phone call, a RFI should be used. RFIs can be used to document verbal clarifications.

Companies often have a standard RFI form. The construction manager should track and follow up on RIFs to insure that the required information is received in a timely manner. An RFI should contain the following information:

- Project name and number
- Unique RFI number used for tracking
- Date of the RFI
- To whom the RFI is being sent
- Who is sending the RFI
- Who initiated the RFI
- Description of the requested information
- Date when response to the RFI is needed
- Signatures

Change Orders

All changes in the scope of work should be documented with a written change order, even if they do not result in a change in cost or schedule. The change order modifies the construction contract and acknowledges that the parties to the contract (e.g., the owner and the contractor) are in agreement to the change in project scope, cost, and schedule. The construction manager should track and follow up on all change orders to insure that they are signed in a timely manner and do not delay the project.

Document Tracking

Submittals, RFIs, change orders, subcontracts, and other such communications should be tracked to insure that they are responded to in a timely manner. This is done by keeping a log of the important dates in the life of the document. For example, a submittal log kept by the contractor would track the following dates:

- Date requested from subcontractor
- Date due from subcontractor
- Date received by subcontractor
- Date sent to design professional
- Date due from design professional
- Date received from design professional
- Date returned to subcontractor

The construction manager should follow up to make sure that the due dates are met.

Logs and Daily Report

The construction manager is responsible for maintaining numerous logs including a log of phone calls, list of site visitors, accident reports, log of photographs and video recordings, purchase order log, etc. A daily report should be kept for each job. The daily report should include the following:

- Date
- Project name and number
- Who prepared the report
- Weather conditions
- Contractor's and subcontractors' labor at the jobsite
- Equipment at the jobsite
- Material delivers received
- Visitors to the site

Chapter 5: Building Codes

All building construction is governed by a building code, even if a building permit is not required. States, counties, and cities may adopt (with or without modification) a nationally recognized set of codes or they may write their own. The most commonly adopted codes are the International Family of Codes, written by the International Code Council, and the *National Electrical Code* (NFPA 70), written by the National Fire Protection Association. Care must be taken to design and construct buildings in accordance with the adopted codes. Buildings codes are established to protect the building and its occupants.

International Building Code (IBC)

The IBC covers the design and construction of all buildings except detached one- and two-family residences and townhomes less than three-stories above grade that have separate entrances (IBC 101.2). Buildings built under the IBC must also comply with many of the other building codes, such as the *National Electrical Code*. The IBC has different provisions for different buildings based upon their use (occupancy) and the building's type of construction, which are discussed later in this chapter.

International Residential Code (IRC)

The IRC covers the design and construction of residential structures not covered by the IBC. Unlike the IBC, the IRC includes all mechanical, plumbing, and electrical codes that apply to these residential structures.

International Fire Code (IFC)

The IFC sets standards for fire safety in buildings.

International Fuel Gas Code (IFGC)

The IFGC covers the design and installation of gas lines from the point of delivery to the inlet of the appliance.

International Mechanical Code (IMC)

The IMC covers the design and construction of the mechanical system, including heating, ventilation, air conditioning, and refrigeration systems.

International Plumbing Code (IPC)

The IPC covers the design and construction of the plumbing system including water supply, waste disposal, and medical gasses. The IPC covers both the fixtures and the piping.

International Private Sewage Disposal Code (IPSDC)

The IPSDC covers the design and construction of private sewage disposal facilities such as a septic tank.

National Electrical Code (NFPA 70)

The National Electrical Code covers the design and construction of electrical systems.

International Property Maintenance Code (IPMC)

The IPMC covers the operation and maintenance of existing buildings.

International Energy Conservation Code (IECC)

The IECC sets standards for energy efficiency in buildings.

Common Reference Standards

Building codes and construction documents often reference common standards rather than incorporating them in the text of the document. The following are some of the more common groups that publish standards:

AASHTO	American Association of State Highway and Transportation Officials
ACI	American Concrete Institute
AF&PA	American Forest and Paper Association
AISC	American Institute of Steel Construction
AITC	American Institute of Timber Construction
ANSI	American National Standards Institute
APA	APA – Engineered Wood Association (formerly the American Plywood Association)
ASCE	American Society of Civil Engineers
ASME	American Society of Mechanical Engineers
ASTM	ASTM International (formerly the American Society for Testing and Materials)
AWPA	American Wood Protection Association
AWS	American Welding Society
FM	Factory Mutual Global Research
GA	Gypsum Association
HPVA	Hardwood Plywood Veneer Association
ICC	International Code Council
ISO	International Organization for Standardization
NCMA	National Concrete Masonry Association
NFPA	National Fire Protection Association
PCI	Precast Prestressed Concrete Institute
SDI	Steel Deck Institute
SJI	Steel Joist Institute
UL	Underwriters Laboratories, Inc.

Building Occupancies

The IBC divides the use or occupancy of a building into ten major groups, which may be further subdivided. A building may consist of a single occupancy or have multiple occupancies. The occupancy groups are based upon people related hazards (e.g., number of people, their ages, and their mobility) and building content related hazards (e.g., flammable materials). The occupancy groups are as follows:

Group A: Assembly (IBC 303)
This group consists of occupancies where large groups of people gather. It is subdivided into the follow occupancies:

A-1: Theaters and concert halls that typically have fixed seating.

A-2: Eating and drinking establishments, such as restaurants, night clubs, and bars.

A-3: Recreational facilities, places of worship, gym and pool facilities (without spectator seating), exhibition halls, libraries, museums, and transportation terminals. This occupancy covers any type of assembly not covered in A-1, A-2, A-4, or A-5.

A-4: Rooms where people congregate to watch indoor sporting events.

A-5: Structures where people congregate to watch outdoor sporting events.

Group B: Business (IBC 304)

This group consists of occupancies where business is conducted (except retail) and includes banks, barber shops, college educational facilities, laboratories, showroom, post offices, and professional offices.

Group E: Educational (IBC 305)

This group consists of occupancies used for the education of children in Kindergarten through 12th grade. It may also include some preschools.

Group F: Factory (IBC 306)

This group consists of occupancies used for manufacturing. It is subdivided into two occupancies: **F-1** for moderate-hazard (e.g., automobiles, clothing, food, millwork) and **F-2** for low-hazard (e.g., ice, glass). The manufacturing of high-hazard products is classified as a Group H occupancy.

Group H: High-Hazard (IBC 307)

This group consists of occupancies that have a high-hazard potential due to the materials stored or used in the building. A building is classified as a Group H occupancy not because of the presence of high-hazard material, but the presence of high-hazard materials in excess of the quantity established the IBC. This group is subdivided into five occupancies (**H-1** to **H-5**) based upon the materials in the building, with H-1 being the most hazardous.

Group I: Institutional (IBC 308)
This group consists of buildings where the residents are cared for by a staff. This occupancy is subdivided into four occupancies based upon the occupant's ability to exit the building during an emergency. The occupancy groups are as follows:

I-1: Assisted living centers, rehab centers, and other facilities where the occupants can exit the building without help.

I-2: Nursing homes, hospitals, and child care facilities, where the occupants need help to exit the building.

I-3: Jails and prisons where the occupants are detained.

I-4: Day care facilities.

Group M: Mercantile (IBC 309)
This group consists of retail stores including grocery stores, department stores, and gas stations.

Group R: Residential (IBC 310)
This group consists of buildings where people live, which are not covered by the IRC. It is subdivided into the following four occupancies:

R-1: Hotels, motels, and boarding houses where the occupants are transient (stay a short time).

R-2: Apartments, boarding houses, dormitories, motels, monasteries where the occupants stay a long time.

R-3: All residential occupancies governed by the IBC which are not classified as R-1, R-2, or R-4.

R-4: Assisted living facilities for 5 to 16 persons.

Group S: Storage (IBC 311)

This group consists of occupancies used for storage. It is subdivided into two occupancies: **S-1** for moderate-hazard (e.g., books, clothing, furniture, lumber) and **S-2** for low-hazard (e.g., glass, meats, metals). The storage of high-hazard products is classified as a Group H occupancy.

Group U: Utility and Miscellaneous (IBC 312)

This covers an odd assortment of occupancies not covered elsewhere, including barns, carports, fences (over 6 feet), retaining walls, sheds, tanks, and towers.

Mixed Use and Occupancy

Often buildings contain multiple uses and occupancies. A good example of this is a Nevada hotel, which may contain a fixed-seat theater for shows (Group A-1), restaurants (Group A-2), bars (Group A-2), gaming areas (Group A-3), exhibition halls (Group A-3), shops (Group M), sleeping rooms (Group R-1), and a parking garage (Group S-2). The IBC regulates the fire rating that must be maintained between these occupancies.

Types of Construction

The IBC divides the types of building construction into nine building types. They are as follows:

Type I and II

Type I and II buildings are constructed of primarily noncombustible materials. Combustible materials may be used in limited places (e.g., wood may be used for doors and blocking for handrails). Noncombustible buildings are divided into four types based upon the required fire-resistant rating of the structural frame, walls, floor, and roof. The four types are as follows:

Type I-A: 3-hour rated structural frame and bearing walls.
Type I-B: 2-hour rated structural frame and bearing walls.
Type II-A: 1-hour rated structural frame and bearing walls.
Type II-B: Non-rated structural frame and bearing walls.

Type III

Type III buildings consist of a noncombustible exterior and combustible interior. Type III buildings are divided into two types (**Type III-A** and **Type III-B**). Type III-A has a higher fire-resistant rating than a Type III-B.

Type IV

Type IV buildings are constructed of heavy timber.

Type V

Type V buildings are constructed of any materials allowed by code. Type V includes wood framed buildings. Type V is divided into two types (Type V-A and V-B). Many components of a **Type V-A** have a one-hour fire rating, whereas a **Type V-B** is a non-rated building.

Chapter 6: Overview of Estimating

In order to prepare an accurate estimate, the amount of materials, labor, and equipment needs to be accurately estimated. The estimating of material quantities is covered in Chapter 9, where construction materials are discussed. This chapter covers how to calculate hourly rates for labor and equipment, how to estimate labor productivity, and other general estimating concepts.

Calculating Hourly Burdened Wage Rates

The hourly burdened wage rate includes the wages paid to the employee plus labor burden. The hourly burdened wage rate should be calculated for each employee class (e.g., lead carpenter, carpenter, carpenter's apprentice, plumber) that is to be used on the job.

Wages

Wages are the amount of money paid to the employees in exchange for their work on the project. Gross wages are calculated before deducting taxes, insurance premiums, etc. from the paycheck and include cash equivalents and allowances. The wage rate for each employee class may be determined in one of three ways.

First, it may be set by the contract documents. On projects which are federally funded, the contractor is required to comply with the Davis-Bacon Act and pay at least the prevailing wages set by the general decision assigned to the project. General decisions for specific locations may be downloaded at www.wdol.gov. If a general decision is included in the contract documents, its wage rates should be used rather than any downloaded general decision.

Cities and states may also set minimum wage rates. These are minimum wage rates and the contractor may have to pay a higher rate for the work.

Second, the wages may be set by the union. These rates are set forth in a contract between the union and the contractor or the union and a group of contractors, such as a local trade group.

Third, the wages may be determined by the prevailing market rate, which is the wage rate that the contractor must pay to retain employees given the current labor market conditions. If the prevailing market rates are higher than the Davis-Bacon or union rates, the contractor must pay the prevailing market rate to keep employees.

Most hourly employees must be paid overtime. Typically, overtime is time and a half (1.5 × base wage rate) for any hours worked over 40 hours per week. Union contracts may have different overtime rates for working Sundays and holidays. If overtime is to be worked on the project, the number of overtime hours must be estimated.

Any anticipated bonuses or increases in the wage rate must be included in the wage calculations.

Example: Determine the annual wages for plumbers. The Davis-Bacon Act requires that they get paid $35 per hour. Historically they are paid for 2,080 hours of regular time and 250 hours of overtime, at time and a half, each year.

$$\text{Wages} = 2{,}080 \text{ hr/yr} \times \$35/\text{hr} + 250 \text{ hr/yr} \times 1.5 \times \$35/\text{hr}$$
$$\text{Wages} = \$85{,}925/\text{yr}$$

Labor Burden

Labor burden includes all costs paid by the employer in addition to the base wages. Labor burden does not include costs deducted from the employee's paycheck. Labor burden includes the following costs:

The Davis-Bacon Act often sets a minimum dollar amount for employee benefits. In lieu of providing these benefits, the employer may add to the employee's wages the **Cash Equivalent** (i.e., the required dollar amount) of these benefits. Cash equivalents are treated as wages when calculating other burden items such as taxes.

Example: The Davis-Bacon Act requires that the plumbers in the above example be provided with $3.52 per hour for benefits. What is the cost if the employer decides to pay the cash equivalent rather than providing the benefits?

$$\text{Cash equivalent} = (2{,}080 \text{ hr/yr} + 250 \text{ hr/yr})(\$3.52/\text{hr})$$
$$\text{Cash equivalent} = \$8{,}202/\text{yr}$$

The employer may pay the employee an **Allowance** (a set amount) for use of the employee's personal vehicle or tools at work or for the employee to purchase any required tools or uniforms. Allowances are treated as wages when calculating other burden items. These are not reimbursements for actual costs. Reimbursements are not burden costs.

Example: An employer gives its plumbers a $50 per week gas allowance. What is the annual cost of this allowance?

$$\text{Allowance} = 52 \text{ weeks/yr} \times \$50/\text{week} = \$2{,}600/\text{yr}$$

Employers are required to pay **Payroll Taxes** (social security and Medicare) on the wages paid to employees. For 2010, the social security rate was 6.2% on the first $106,800 of wages for each employee. The rate of 6.2% has remained unchanged for many years, but the amount of wages that social security is paid on changes almost every year. The current rate can be found in *Publication 15 (Circular E) Employer's Tax Guide* which can be downloaded from www.irs.gov. The 2010 Medicare rate was 1.45% of the employee's gross wages. The Medicare rate has remained unchanged for years. The taxes must be paid on wages, cash equivalent, and allowances.

In addition to the employer paying social security and Medicare taxes, the employee must also pay social security and Medicare taxes. The employees' rate is the same as the employer's and is deducted from their wages. The employee's portion of these taxes is not part of the labor burden.

Example: A plumber is paid $96,727 per year, which includes wages, cash equivalents, and allowances. Using the above rates, determine the labor burden cost of social security and Medicare taxes.

Only the employer's position of these taxes is part of the labor burden. These costs are calculated as follows:

Social security = $96,727/yr $\times 0.062 = \$5,997/yr$

Medicare = $96,727/yr $\times 0.0145 = \$1,403/yr$

Employers are required to provide **Unemployment Insurance** for their employees. For 2010, the federal unemployment (FUTA) rate was 6.2% on the first $7,000 of each employee's wages. The employer may deduct 5.4% on the first $7,000 of each employee's wages for employees covered by state unemployment insurance (SUTA), provided the state payments are made on time. When the 5.4% deduction is taken, the federal unemployment rate becomes 0.8% on the first $7,000 of wages. The current rate can be found in *Publication 15 (Circular E) Employer's Tax Guide*. The SUTA rate varies by state and by employer. The employer's rate is based on unemployment claims made against the employer.

Example: Using the above FUTA rate and a SUTA rate of 2.6% on $20,000 of wages, determine the cost of unemployment insurance for the plumber in the above example. The employer is eligible to take the maximum deduction.

The SUTA is calculated as follows:

$$SUTA = \$20,000/yr \times 0.026 = \$520/yr$$

The FUTA rate is 0.8% on the first $7,000 of wages. The FUTA is calculated as follows:

$$FUTA = \$7,000/yr \times 0.008 = \$56/yr$$

Employers are required to provide **Workers' Compensation Insurance**, which covers the employee if they are injured on the job. The cost of the insurance varies by state, company, and job classification. Companies with a higher claim history pay a higher insurance rate. A higher rate is also paid on jobs (e.g., roofing) that have higher accident rates and claim histories. Lower rates are paid on jobs (e.g., supervisory work) that have low accident rates and claims histories. Workers' compensation insurance is expressed as dollars per $100 of gross payroll, which includes wages, cash equivalents, and allowances paid to the employee.

Example: The workers' compensation rate for the plumber in the above example is $8.45 per $100 of wages. What is the cost of the workers' compensation insurance?

$$\text{WC insurance} = \$8.45 \times \frac{\$96,727/\text{yr}}{\$100} = \$8,173/\text{yr}$$

General Liability Insurance protects the company against claims due to negligent business activities and the employee's failure to use reasonable care. It covers losses such as bodily injury, property damage or loss, and slander. The insurance premiums are based on the employee's gross wages and the premium rate is higher for management employees.

Example: The general liability insurance rate for the plumber in the above example is 0.75% of gross wages. What is the cost of the general liability insurance?

$$\text{General liability} = \$96,727/\text{yr} \times 0.0075 = \$725/\text{yr}$$

Insurance Benefits include health, dental, and life insurance provided to the employees as part of their benefits packages. Often the employee pays part of the cost of these benefits. Only the costs borne by the employer are part of the burden. The costs paid by the employee are deducted from their gross wages.

Example: Health insurance for the plumber in the above examples costs $600 per month. The employer pays 2/3 of the cost and the employee pays 1/3 of the cost. What is the labor burden cost of the health insurance?

$$\text{Health insurance} = \$600/\text{mo} \times \frac{2}{3} \times 12 \text{ mo/yr} = \$4,800/\text{yr}$$

An employer may provide **Retirement Contributions** to an employee's retirement plan, such as a 401(k), as part of the benefits package. The employer's contributions to a retirement plan are part of the labor burden. The employees' contributions are deducted from their wages and are not considered burden.

Example: The plumber in the above examples may deposit up to 10% of his wages in a company sponsored 401(k) plan. The employer will contribute $0.50 for every $1.00 the plumber contributes. The maximum the employer will contribute is 3% of the employee's wages. Assuming that the employee takes full advantage of the retirement benefits, what is the labor burden cost of the retirement?

$$\text{Retirement} = \$96,727/\text{yr} \times 0.03 = \$2,902/\text{yr}$$

Union Payments are payments to the union to provide benefits (i.e., health insurance and pension plan) and training to union employees. The union payments are specified in the contract with the union. The contract may also require that the employer deduct union dues from the employee's paycheck. Dues deducted from the employee's paycheck are not part of the labor burden.

Example: The plumber in the above example belongs to the local union. The union contract requires that the employer pay the union 1% of the employees' wages to provide training and requires the company to collect 2% of the employees' wages for union dues. What is the labor burden cost of the union payments?

The portion paid by the employee is not included.

$$\text{Union payments} = \$96,727/\text{yr} \times 0.01 = \$967/\text{yr}$$

Employers may provide **Paid Leave** (e.g., vacation, sick leave) for their employees. The employer must pay social security taxes, Medicare taxes, and other burden cost on paid leave. The easiest way to include paid leave in the burden is to include paid leave in the wages and deduct the paid leave hours from the billable hours.

Example: The plumber in the above examples is given four weeks (40 hours per week) of paid vacation and sick leave. Assuming the rest of the hours are billable, how many billable hours does the plumber work a year?

$$\text{Billable hours} = 2,080 \text{ hr} + 250 \text{ hr} - 4 \text{ wk} \times 40 \text{ hr/wk}$$
$$\text{Billable hours} = 2,170 \text{ hr/yr}$$

Hourly Burdened Wage Rates

The hourly burdened wage rate is calculated by determining the annual cost (both wages and burden) for an employee and dividing it by the estimated number of hours that the employee can be billed to projects during the year using the following equation:

$$\text{Burdened wage rate} = \frac{\text{Annual costs}}{\text{Billable hours}}$$

Example: Determine the hourly burdened wage rate for the plumber in the above examples.

$$\text{Wages} = \$85,925/\text{yr}$$
$$\text{Cash equivalent} = \$8,202/\text{yr}$$
$$\text{Allowance} = \$2,600/\text{yr}$$
$$\text{Social security} = \$5,997/\text{yr}$$
$$\text{Medicare} = \$1,403/\text{yr}$$
$$\text{SUTA} = \$520/\text{yr}$$
$$\text{FUTA} = \$56/\text{yr}$$
$$\text{WC insurance} = \$8,173/\text{yr}$$
$$\text{General liability} = \$725/\text{yr}$$
$$\text{Health insurance} = \$4,800/\text{yr}$$
$$\text{Retirement} = \$2,902/\text{yr}$$
$$\text{Union payments} = \$967/\text{yr}$$
$$\text{Total} = \$122,270$$

The hourly burdened wage rate is calculated as follows using the 2,170 hours from the above example:

$$\text{Burdened Wage Rate} = \frac{\$122,270}{2,170\ \text{hr/yr}} = \$56.35/\text{hr}$$

Equipment Costs

Equipment costs may be divided into two broad categories: ownership and operating costs. Ownership costs are those costs which are nearly fixed and are a result of simply owning a piece of equipment. Ownership costs include depreciation and interest, taxes and licensing, insurance, and storage. Operating costs are those costs which vary with the use of the equipment and include tires and other wear items, fuel, lubricants and filters, and repair reserves (money set aside for repairs).

Depreciation and Interest

Depreciation and interest cover the loss in value for a piece of equipment as it ages and a return on the capital tied up in its purchase (or interest on its loans). Its annual cost is calculated by using the following equation.

$$\text{Cost}_{\text{D\&I}} = \left[\frac{Pi(1+i)^n}{(1+i)^n - 1} \right] - \left[\frac{Fi}{(1+i)^n - 1} \right]$$

where

    P = Purchase price, including sales tax, transportation, set up, etc., and excluding the cost of tires and other major wear items, which are treated as an operating cost

    F = Salvage value or the expected sale price at the end of its useful life

    i = Interest rate expressed in decimal format

    n = Useful life of the equipment in years

Example: Determine the depreciation and interest cost for a $200,000 front-end loader using an interest rate of 10%. The price includes all purchase costs and the tires. A set of tires for the loader cost $32,000. The estimated salvage value at the end of its six year life is $30,000. The loader is expected to be used 1,500 hours per year.

The tires must be deducted from the purchase price. The purchase price is $168,000 ($200,000 - $32,000).

$$\text{Cost}_{D\&I} = \left[\frac{\$168,000(0.1)(1+0.1)^6}{(1+0.1)^6 - 1} \right] - \left[\frac{\$30,000(0.1)}{(1+0.1)^6 - 1} \right]$$

$$\text{Cost}_{D\&I} = \frac{\$34,686/\text{yr}}{1,500 \text{ hr/yr}} = \$23.12/\text{hr}$$

Taxes and Licensing
For vehicles that travel on public roads, the property taxes and licensing fees are paid when the vehicle is registered. Property tax is paid on off-highway vehicles and heavy equipment annually with the company's state income tax.

Example: The property tax for the loader in the above example is 1.2% of its valuation and there is no licensing cost. For the second year it is valued at 56% of its purchase price. What is property tax for the second year?

$$\text{Valuation} = \$200,000 \times 0.56 = \$112,000$$

$$\text{Cost}_{\text{T\&L}} = \$112,000 \times 0.012 = \frac{\$1,344/\text{yr}}{1,500 \text{ hr/yr}} = \$0.90/\text{hr}$$

Insurance
Insurance covers damage to the equipment and damaged caused by the equipment. The cost of the insurance is obtained from the company's insurance policy or agent.

Storage
The cost for storage and shop facilities used to maintain the equipment is allocated to the equipment. This is done by totaling the costs a dividing them among the various pieces of equipment.

Ownership Cost

Example: The annual insurance cost for the loader in the above examples is $3,000. The company allocated $2,000 to the loader to cover the cost of storage. What is the hourly ownership cost for the loader?

$$\text{Cost}_{\text{Insurance}} = \frac{\$3000/\text{yr}}{1{,}500\ \text{hr/yr}} = \$2.00/\text{hr}$$

$$\text{Cost}_{\text{Storage}} = \frac{\$2{,}000/\text{yr}}{1{,}500\ \text{hr/yr}} = \$1.33/\text{hr}$$

$$\text{Cost}_{\text{Ownership}} = \$23.12/\text{hr} + \$0.90/\text{hr} + \$2.00/\text{hr} + 1.33/\text{hr}$$

$$\text{Cost}_{\text{Ownership}} = \$27.35/\text{hr}$$

Tires and Other Wear Items

For tires and wear items that have a useful life of less than one year, their operating cost is their cost divided by their useful life. For tires and wear items that have a useful life of more than one year, their operating cost is calculated in the same manner as depreciation and interests cost for a piece of equipment using the useful life of the wear item and a salvage value of zero. The costs are calculated using the following equation:

$$\text{Cost}_{\text{Tires}} = \left[\frac{Pi(1+i)^n}{(1+i)^n - 1} \right]$$

The cost of tire repair must be added to the above costs. The cost of tire repair is calculated using the following equation. A common repair multiplier is 15%.

$$\text{Cost}_{\text{Tire Repair}} = \frac{\text{Repair multiplyer} \times \text{Tire cost}}{\text{Tire life}}$$

Example: The life of the tires for the loader in the above examples is 3,000 hours. Determine the hourly cost of tires and tire repair using a repair multiplier of 15%.

$$\text{Tire life} = \frac{3,000 \text{ hr}}{1,500 \text{ hr/yr}} = 2 \text{ yr}$$

$$\text{Cost}_{\text{Tires}} = \left[\frac{\$32,000(0.1)(1+0.1)^2}{(1+0.1)^2 - 1} \right] = \$18,438/\text{yr}$$

$$\text{Cost}_{\text{Tires}} = \frac{\$18,438/\text{yr}}{1,500 \text{ hr/yr}} = \$12.29/\text{hr}$$

$$\text{Cost}_{\text{Tire Repair}} = \frac{0.15 \times \$32,000}{3,000 \text{ hr}} = \$1.60/\text{hr}$$

Fuel

The fuel cost is calculated by multiplying the hourly fuel consumption by the cost per gallon of fuel. The fuel consumption varies depending on the type of work the equipment is doing. An excavator digging in hard clay will consume more fuel than one digging in sand. The best source for estimated fuel consumption is from historical data or the equipment's manufacturer. Alternately, the fuel consumption can be estimated by the following formula.

$$\text{Fuel}_{\text{Gas}} = 0.06 \times \text{hp} \times \text{Power utilization} \times \text{Use factor}$$

$$\text{Fuel}_{\text{Diesel}} = 0.04 \times \text{hp} \times \text{Power utilization} \times \text{Use factor}$$

where
   hp = Horse power rating of the engine
   Power utilization = Percent of available engine power
      that is being used, typically 55% to 80%.
   Use factor = Percent of an hour that the equipment is
      being used, typically between 75% (45 min/hr) and
      83% (50 min/hr)

Example: The loader in the above examples has a 250-hp diesel engine. Estimate the hourly fuel cost using a power utilization of 70%, a use factor of 75%, and a fuel cost of $3.25 per gallon.

$$\text{Fuel}_{\text{Diesel}} = 0.04 \times 250 \text{ hp} \times 0.70 \times 0.75 = 5.25 \text{ gal/hr}$$
$$\text{Cost}_{\text{Fuel}} = 5.25 \text{ gal/hr} \times \$3.25/\text{gal} = \$17.06/\text{hr}$$

Lubricants and Filters
Lubricants and filters include the costs of lubricants, filters, and the service technician needed to complete the scheduled maintenance. Lubricants and filters also include the cost of the lubricants used at the beginning of each shift, typically grease for the joints. The frequency of the scheduled maintenance varies based on the working conditions. Trucks operating on the highway require less frequent maintenance than off-highway trucks, which are operating in a dusty environment. The maintenance requirements are obtained from the manufacturer.

Example: The lubricant consumption (from the manufacturer) and costs for the loader in the above examples is shown in the table below. The filter costs for the loader are $0.18/hr and a tube of grease that costs $4.24 is used at the beginning of each shift. What are the lubricant and filter costs for the loader?

Use	gal/hr	$/gal
Crankcase (oil)	0.010	7.43
Transmission	0.006	5.83
Drives & differential	0.006	9.34
Hydraulic controls	0.007	3.61

$$\text{Cost}_{L\&F} = (0.010 \text{ gal/hr})(\$7.43/\text{gal})$$
$$+ (0.006 \text{ gal/hr})(\$5.83/\text{gal})$$
$$+ (0.006 \text{ gal/hr})(\$9.34/\text{gal})$$
$$+ (0.007 \text{ gal/hr})(\$3.61/\text{gal})$$
$$+ (1 \text{ tube/8 hr})(\$4.24/\text{tube}) + \$0.18/\text{hr}$$
$$\text{Cost}_{L\&F} = \$0.90/\text{hr}$$

Repair Reserve
The repair reserve is the amount of money that needs to be set aside for each hour of operation for future repairs and overhauls. This information can be obtained from the manufacturer.

Operating Costs
Example: The estimated repair reserve for the loader in the above examples is $8.50 per hour. Determine the total operating cost for the loader. What is the total hourly cost?

$$\text{Cost}_{\text{Operating}} = \$12.29/\text{hr} + \$1.60/\text{hr} + \$17.06/\text{hr}$$
$$+ \$0.90/\text{hr} + \$8.50/\text{hr}$$
$$\text{Cost}_{\text{Operating}} = \$40.35/\text{hr}$$
$$\text{Cost}_{\text{Total}} = \$40.35/\text{hr} + \$27.35/\text{hr} = \$67.70/\text{hr}$$

Leased and Rented Equipment
When leasing or renting equipment some of the ownership and operating costs are covered by the lease (rental) payments. To determine the equipment cost for leased (rented) equipment, the ownership and operating costs that are not covered by the lease (rental) agreement are added to the lease (rental) cost.

Example: The loader in the above examples can be rented for $55.00 per hour. The rent covers all cost except fuel. What is the cost to rent the loader?

$$\text{Cost}_{\text{Total}} = \$55.00/\text{hr} + \$17.06/\text{hr} = \$72.06/\text{hr}$$

Crew Rates

Because construction crews are made up of different classes of workers getting paid different rates, a crew rate that takes the crew mix into account must be calculated. The crew rate may be expressed as the cost for the entire crew for one clock hour (or day) or it may be expressed as the average labor cost per labor hour. The crew rate per hour (or day) is calculated by summing the labor costs for the crew for the hour (or day). The crew rate per labor hour (lhr) is calculated by dividing the crew rate per hour (or day) by the number of labor hours in one hour (or day); and is calculated using the following equation:

$$\text{Labor cost/lhr} = \frac{\text{Total labor cost}}{\text{Labor hours}}$$

Example: Determine the labor cost per clock hour and per labor hour for a paving crew. The crew consists of a foreperson ($52.21), four paver operators ($40.10), three roller operators ($35.17), and one laborer ($30.75). The burdened hourly labor rates are in the parentheses.

$$\text{Labor cost/hr} = 1 \times \$52.21/\text{hr} + 4 \times \$40.10/\text{hr}$$
$$+ 3 \times \$35.17/\text{hr} + 1 \times \$30.75/\text{hr}$$

$$\text{Labor cost/hr} = \$348.87/\text{hr}$$

$$\text{Labor hr} = 1 + 4 + 3 + 1 = 9 \text{ lhr per hr}$$

$$\text{Labor cost/lhr} = \frac{\$348.87/\text{hr}}{9 \text{ lhr/hr}} = \$38.76/\text{hr}$$

Because productivity is measured in output per crew hour (or day) or in labor hours required to produce one unit of work, the equipment cost must be expressed in cost per crew hour (day) or average equipment cost per labor hour. The equipment cost per hour (or day) is calculated by summing the hourly (or daily) equipment costs for the hour (or day). The equipment cost per labor hour is calculated by dividing the equipment costs per hour (or day) by the number of labor hours in one hour (or day). It is important to note that it is divided by the labor hours, **not** the number of pieces of equipment. The equipment cost per labor hour is calculated by the following equation:

$$\text{Equipment cost/lhr} = \frac{\text{Total equipment cost}}{\text{Labor hours}}$$

Example: Determine the equipment cost per clock hour and per labor hour for a paving crew in the above example. The equipment consists of a pickup truck ($19.52), an asphalt paver ($149.50), a shuttle buggy ($275.25), two Cat CS563 compactors ($45.04), and a Cat CB634 compactor ($69.41). The hourly equipment costs are in the parentheses.

$$\text{Equipment cost/hr} = \$19.52/\text{hr} + \$149.50/\text{hr}$$
$$+ \$275.25/\text{hr} + 2 \times \$45.04/\text{hr} + \$69.41/\text{hr}$$
$$\text{Equipment cost/hr} = \$603.76/\text{hr}$$
$$\text{Equipment cost/lhr} = \frac{\$603.76/\text{hr}}{9 \text{ lhr/hr}} = \$67.08/\text{hr}$$

Productivity

Productivity is a measure of how quickly construction tasks or activities can be performed. The two most common measures of productivity are the number of labor hours it takes to produce one unit of work (e.g., 0.1 labor hours per square foot) and the number of units of work that can be completed by a crew in one day (e.g., 400 square feet per day). A labor hour is defined as one person working for one hour. The best source for determining productivity is from historical data. The labor hours per unit of work can be calculated by the following equation:

$$\text{Labor-hours/Unit} = \frac{\text{Labor-hours}}{\text{Quantity}}$$

Example: A recent project included 2,540 square feet of 4-inch-thick sidewalk. It took 22 labor-hours to complete the sidewalk. Determine the productivity in labor-hours per square foot based on this historical data.

$$\text{Labor-hours/Unit} = \frac{22 \text{ lhr}}{2,540 \text{ ft}^2} = 0.0087 \text{ lhr/ft}^2$$

The crew output, in units per crew day, can be calculated by the following equation:

$$\text{Output} = \frac{\text{Quantity}}{\text{Crew days}}$$

Example: A recent project included 1,950 square feet of 8-inch block wall. It took five crew days to complete the wall. Determine the crew output in units per crew day.

$$\text{Output} = \frac{1,950 \text{ ft}^2}{5 \text{ crew days}} = 390 \text{ ft}^2 \text{ per crew day}$$

Factors Affecting Productivity

Many factors affect productivity. These factors include:

Project Size. Larger projects tend to have higher productivity rates because crews can become more specialized and they have the chance to progress down the learning curve.

Overtime. Working overtime decreases the productivity. The longer this goes on, the greater the reduction.

Size of Crew. If a crew is too large the workers get in each other's way, reducing productivity. If a crew is too small, they often have too few people to efficiently complete a task (e.g., standing up a wall), which reduces productivity.

Delays. Delays in getting materials and equipment or delays in getting the project ready for the crews decrease production. This is one of the biggest enemies to efficiency.

Interruptions. Anytime a task is interrupted the productivity decreases.

Weather. Working in adverse weather reduces productivity. Weather that is too hot or too cold slows workers down as they try to maintain body temperature. Rain or snow slows down work and may require removal (shoveling or pumping) before work can continue.

Project Layout. The location of the materials storage, toilets, etc. has a great effect on productivity. As the distance to these increases, the amount of time spent moving about the site increases and the time spent installing materials decreases; reducing production.

Safety. The use of safety equipment decreases production. A task that can be performed quickly on the ground will be less productive high up on a building where additional safety measures are needed. Safety should always be more important than production.

Adjustments need to be made to the historical productivity rates. This is done by multiplying the historical productivity rate by an adjustment factor (AF).

When working with labor hours per unit of work, an adjustment factor less than one increases productivity and an adjustment factor greater than one decreases productivity.

When working with output (units per crew hour), an adjustment factor less than one decreases productivity and an adjustment factor greater than one increases productivity.

Cycle Time

A common way to determine productivity is to perform a cycle time analysis. Cycle time analysis works for cyclical activities, such as trucks hauling dirt from an excavation site to a disposal site.

The first step in performing a cycle time analysis is to determine the average cycle time. This is done by measuring a number of cycles (preferably at least 30 at different times during the day) and determining their average using the following equation:

$$CT_{Ave} = \frac{(CT_1 + CT_2 + \cdots + CT_n)}{n}$$

where

CT = Cycle time

n = Number of observations

Next, using the average cycle time, the productivity is calculated. The following equation is used to calculate the productivity in labor hour per unit:

$$\text{Labor-hours/Unit} = \frac{(CT_{Ave})(AF)(Size)}{(SE)(Units)}$$

where

CT_{Ave} = Average cycle time in minutes per cycle

AF = Adjustment factor

Size = Size of crew needed to complete the cycle

SE = System efficiency (minutes per 60-minute hour), which takes into account wait and other nonproductive time, typically 45 or 50 min/hr

Units = Number of units produced per cycle

In this equation, an adjustment factor less than one increases productivity and an adjustment factor greater than one decreases productivity.

Example: A tractor and trailer is used to haul asphalt from the asphalt plant to a road paving project. Three cycle times have been measured at three different times during the day. The cycle times were 120 minutes, 108 minutes, and 112 minutes. The tractor and trailer requires one operator and can haul 20 tons of asphalt per cycle. Using a system efficiency of 45 minutes per hour and an adjustment factor of 1.1, determine the labor productivity for hauling asphalt.

$$CT_{Ave} = \frac{(120 \text{ min} + 108 \text{ min} + 112 \text{ min})}{3} = 113 \text{ min}$$

$$\text{Labor-hours/Unit} = \frac{(113 \text{ min})(1.1)(1 \text{ lhr/hr})}{(45 \text{ min/hr})(20 \text{ tons})}$$

$$\text{Labor-hours/Unit} = 0.138 \text{ lhr/ton}$$

Rate of Progress

Another way to determine productivity is rate of progress. Rate of progress is used for linear tasks, such as paving a road. Rate of progress is determined using the following equation:

$$\text{Lhr/Unit} = \frac{\left(\dfrac{\text{Quantity}}{\text{RoP}} + \text{TT}\right)(\text{AF})(\text{Size})}{(\text{Quantity})(\text{SE})}$$

where

  Quantity = Quantity of work to be performed
  RoP = Rate of progress: How fast a task can be performed, which is measured in units per minute
  TT = Travel time: Nonproductive time that occurs when a piece of equipment is being moved, which is measured in minutes
  AF = Adjustment factor
  Size = Size of crew
  SE = System efficiency (minutes per 60-minute hour)

Example: A 40,000 square yard parking lot is being paved in fifty 4-yard-wide passes. At the end of each pass it takes 10 minutes to turn the paver around. The paver can pave 16 square yards per minute. Using a system efficiency of 50 minutes per hour and an adjustment factor of 1.05, determine the labor productivity for paving. There are 11 workers on the paving crew.

$$\text{Lhr/Unit} = \frac{\left(\dfrac{40,000 \ \text{yd}^2}{16 \ \text{yd}^2/\text{min}} + 49 \times 10 \ \text{min}\right)(1.05)(11 \ \text{lhr/hr})}{(40,000 \ \text{yd}^2)(50 \ \text{min/hr})}$$

$$\text{Lhr/Unit} = 0.0173 \ \text{lhr/yd}^2$$

Project Overhead (General Conditions) Checklist

The following is a list of items to be sure to include when preparing an overhead budget. Where possible, general labor and rental equipment should be included in the building costs. They are included in the general overhead when they are used on so many different activates that tracking where they are used becomes impracticable.

Project Supervision:
    Project Manager
    Assistant Project Manager
    Superintendent
    Project Engineer
    Project Estimator
    Project Accountant
    Project Secretary
    Detailer/Draftsman

General Labor:
    Labor Foreman
    Laborers

Engineering
    Engineering
    Surveying

Testing and Inspections:

Soils Report
Soil Testing and Inspections
Concrete Testing and Inspections
Rebar Inspections
Masonry Testing and Inspections
Welding Testing and Inspections
Other Testing and Inspections

Safety:

First Aid Kits
Hard Hats
Safety Equipment
Safety Rails, Toe Boards, etc.
Other Safety Supplies

Temporary Offices and Facilities:

Office Trailer
Mobilize/Demobilize Trailer
Trailer Skirts
Utility Hookups for Trailer
Storage Bins

Temporary Utilities:

Temporary Telephone
Temporary Power
Temporary Heating
Temporary Water
Temporary Lights
Temporary Toilets
Internet Connection

Barriers and Enclosures:
Temporary Fencing and Gates
Pedestrian Canopies
Temporary Barricades
Dust Partitions
Weather Protection

Traffic Control:
Flaggers
Barricades
Flares and Lights

Rental Equipment and Tools:
Scaffolding
Personnel Lift
Forklift
Crane
Other Equipment and Tools

Project Supplies:
Office Supplies
Office Furnishings
Fax/Copy Machines
Computers
Software
Drinking Water
Radios
Blue Print Reproduction
Construction Photos
Postage/Federal Express

Security:
   On-site Guards
   Security Patrols
   Security System

Access and Parking:
  Access Road Construction
  Access Road Maintenance
  Temporary Parking
  Easements

Cleanup:
   Cleanup Labor
   Trash Removal and Disposal

Signage:
   Project Signs
   Safety Signs
   Other Signs

Subsistence:
   Temporary Housing
   Maid Service
   Air Fare
   Hotel Rooms
   Rental Car and Fuel
   Meals and Expenses

Final Inspection and Close Out:
   Warranty
   Damage Repairs
   Maintenance Manuals
   Commissioning

Bonds and Insurance:
   Prime Contract Bonds
   Subcontractor Bonds
   Increased Liability Limits
   Builder's Risk

City Fees and Permits:
   Plan Check Fee
   Building Permit Fee
   Grading or Excavation Permit
   Sewer Assessment Fee
   Water Meter Fee
   Storm Drainage Fee
   Power Company Charges
   Gas Company Charges
   Telephone Company Charges
   Fire Water Connection Fee
   Business License
   Encroachment Permit

Standard Markups

Markups are added to the bid after the construction costs and project overhead (indirect overhead) have been calculated. The construction costs include the cost of all labor, materials, equipment, subcontractor work, and other costs that are required to install the components necessary to complete the construction project. The project overhead includes all costs that are required to complete the project, but are not part of installing components in the building. Most MasterFormat Division 1 costs are project overhead costs. Markups may be based upon the construction costs (including project overhead), the total bid, a cost classification (e.g., materials, labor), or some other subtotal.

Sales Tax

Sales tax is often considered a markup, which is a percentage that is added to the cost of the materials. When preparing a project's budget, the sales tax must be spread out among the cost codes that include materials. This is because the sales tax will be billed along with the material costs to the individual cost codes. Sometimes sales tax is included in the materials cost rather than treated as a markup. In some states, sales tax may be charged on all services, making all costs including profit and general overhead, subject to sales tax.

Example: Determine the sales tax on a lumber order with a cost of $117,000. The sales tax rate is 6.75%.

$$\text{Sales tax} = \$117,000 \times 0.065 = \$7,605$$

Building Permits

Many buildings require that the contractor obtain a building permit from the local municipality. The building permit fee covers the cost of building inspections. Federal projects and many municipal projects may be exempt from building permits. Building permits may be based upon the contractor's bid or a set valuation per square foot (e.g. $100 per square foot). The local municipality should be contacted to see if a permit is required and how the cost of the permit will be calculated.

Example: A contractor wants to build a 3,000 square foot home with a 600 square foot garage. The local municipality values the home at $100 per square foot and the garage at $40 per square foot. The cost for the building permit is $993.75 for the first $100,000 and $5.60 for each $1,000 (or fraction thereof) over $100,000.

$$\text{Valuation} = 3,000 \text{ ft}^2 \times \$100/\text{ft}^2 + 600 \text{ ft}^2 \times \$40/\text{ft}^2$$
$$\text{Valuation} = \$324,000$$

The building permit exceeds $100,000 by $224,000 ($324,000 - $100,000) or 224 $1,000s or fractions thereof.

$$\text{Building permit} = \$993.75 + 224 \times \$5.60 = \$2,248$$

Bonds

The owner may require the contractor to provide payment and performance bonds on a project. The contractor may require the major subcontractors (e.g., mechanical, plumbing, electrical) on the project to provide payment and performance bonds. The bonding requirements are found in the project manual. The cost of the bond is based on the total bid for the project, including any alternatives.

Example: Determine the cost of a bond for a project with a total bid of $210,000. The bond rate is 1.5% on the first $50,000, 1.25% on the next $50,000 ($50,001-100,000), and 1% of the remaining amount.

$$Bond = \$50,000 \times 0.015 + \$50,000 \times 0.0125$$
$$+ (\$210,000 - \$100,000)0.01$$
$$Bond = \$2,475$$

Profit and General Overhead

Profit provides a return on investment for the company's owners. Profit is often a percentage of construction costs, although it may be a fixed fee. The profit markup may be less on subcontractor work because there is less risk.

The general overhead markup provides funds to cover the cost of general overhead, overhead that cannot be billed to projects (e.g., main office overhead).

Example: Determine the profit and general overhead markup for a warehouse using an 8% markup. The cost of the warehouse, including all other markups, is $525,000.

$$Profit \ and \ overhead = \$525,000 \times 0.08 = \$42,000$$

Avoiding Estimating Errors

When preparing an estimate, care must be taken to ensure that it is accurate (e.g., has the right quantity of materials, uses appropriated productivity rates) and that it is complete (i.e., includes all of the project's scope). Errors in estimating may result in the company losing the bid or winning an unprofitable job. Errors in material quantities lead to construction delays and increased costs while the correct materials are sent to the jobsite. The following are a few things that can be done to reduce estimating errors.

Spend More Time on Large Costs. A small error on an item with a large cost is often more costly than a large error on an item with a small cost. More time should be spent on the large cost items in the estimates.

Prepare Detailed Estimates. When there is sufficient information in the drawings, a detailed estimate (an estimate that itemizes all of the components) should be prepared. Detailed estimates are more accurate than estimates based upon cost per square foot.

Mark Off Items Counted during the Quantity Takeoff. Marking off items helps ensure that all items are accounted for and items have not been taken off twice. When all items on a detail (or page) are taken off, the detail (or page) should be marked off with a highlighter.

Double-Check All Takeoffs. All takeoff should be performed twice. The adage "Measure twice and cut once" applies to estimating. In the estimator's case it is count twice, order once.

Include Units in Calculations. Including units in your calculations and canceling the units out helps ensure that the correct conversion factors are used.

Automate with Software. The use of spreadsheet and other estimating software speeds up the calculation process and reduces the number of calculation errors provided the software is set up and used correctly. The software should use well-tested and checked formulas. The user must understand how to use the software and its limitations to ensure its proper use. Poorly set up or incorrectly used software can quickly increase the number of estimating errors.

Double-Check All Calculations. Math errors are a common source of estimating errors.

Have Someone Review the Estimate. A second set of eyes will often pick up an error missed by the estimator. The boss or a peer should review all estimates.

Compare the Percentage of Each Cost Code. Similar projects should have similar costs and similar cost breakdown by cost code (e.g., electrical is 15-18% of the building's budget). If a cost looks significantly different from other jobs or is too high (or low) compared to other costs on the project, the cost should be doubled checked.

Allow Plenty of Time. This helps avoid errors made when the estimator is in a rush. Everything that can be done before the bid day should be so that there is plenty of time to complete the estimate.

Chapter 7: Introduction to Construction Management

Job Titles and Roles

The number of people required to manage a project and the organization of the project management team varies based on the size of the construction company and the size of the project. In a small company or on a small project, a manager may fulfill multiple roles. The following are some of the most common management roles in a construction company and their typical duties.

The **Project Manager** is responsible for the success of the project from the time the contractor becomes involved in the project. For design-bid-build projects, the project manager oversees the preparation and submission of the bid; and if the company is the successful bidder, the scheduling, buyout, construction, and close out of the project. For design-build projects and projects where the contractor acts as a construction manager, the project manager also is involved in the design of the project.

The **Estimator** is responsible for preparing the bid for the construction project. The estimator determines the quantity of materials, labor, and equipment needed to complete the project. Multiple estimators may work on a large project; whereas on a small project the estimate may be prepared by the project manager. The estimator may prepare change orders or they may be prepared by field personnel.

The **Purchaser** is responsible for purchasing the bulk of the materials needed for the project based on the material quantities provided by the estimator. In smaller companies, the estimator may also be the purchaser.

The **Scheduler** is responsible for preparing the construction schedule for the project. The schedule is based upon the labor hours and crew sizes used in the estimate. The project manager, estimator, superintendent, other key field personnel, and key subcontractors should have input into the construction schedule.

On larger jobs, **Accounting** and **Secretarial** staff is assigned to the project to providing up-to-date cost data and office support. On smaller projects, these functions are performed by the staff in the contractor's office.

The **Superintendent** is responsible for the day-to-day construction activities at the job site. She oversees the on-site project management team, subcontractors, and the contractor's crews.

The **Project Engineer** is responsible for the paperwork required by the project. He handles change orders, requests for information, submittals and shop drawings, project documentation, and other correspondence.

Field Engineers assist the project engineer and perform quality control and layout.

Forepersons are responsible for overseeing the work of the contractor's crews.

Cost, Time, Quality Relationship

The three primary focuses of a project are time, cost, and quality. **Time** is defined as the length of time it takes to complete a project. **Cost** is defined as the amount of money spent on a project. **Quality** embodies such characteristics as level of finishes (e.g., tile versus vinyl floors), level of workmanship, spaciousness of the rooms, ability to meet the intended use of the project, the environmental impacts of the project (e.g., LEEDs gold), and so forth. The relationship between time, cost, and quality is often displayed on a triangle as shown below.

During the design process, the time, cost, and quality of the project are established. For the designers this becomes a balancing act, because it is not possible to optimize time, cost, and quality individually. As one is optimized one or both of the other two suffer. As the selected quality increases, the cost increases and time often increases. As time decreases below the optimum duration, cost increases and quality often decreases. When balancing time, cost, and quality, the designers must take into account how the owners judge the success of the project. Some may place their emphasis on quality, while others on time or cost. The balance between time, cost, and quality should reflect the priorities of the project owners.

By the time the project is under construction, the standards of time (i.e., the schedule), cost (i.e., the budget), and quality (i.e., the specifications) have been established and the relationship between time, cost, and quality changes. Failure to meet time, cost, or quality usually has a negative impact on the other two factors. Failure to meet the minimum quality standard requires rework, which increases costs and often increases the time it takes to complete the project. Failure to meet the schedule increases costs as money is spent on overtime, expediting material shipments, and other measures to make up time. Quality often suffers as the workers hurry to catch up. For a project to be a success, the project must meet the established standards for time, cost, and quality.

Sustainable (Green) Construction

Sustainable Construction is the practice of designing and building structures that minimize the environmental impact of the structure over its lifetime or life cycle. Sustainable construction considers the total environmental impact of the structure during its construction, operation and maintenance, renovation, and demolition.

Sustainable construction methods are methods that can be used over a long period of time without damaging the environment and may include recycling or reuse of materials, use of easily renewable materials (such as bamboo), and reduction of the carbon footprint of the building.

Additionally, sustainable construction should seek to find new and better ways for humans to interact with buildings such that buildings are used more efficiently and are utilized more fully. For example, university buildings often sit idle during the afternoon and much of the summer. A sustainable practice for a university is to seek to utilize these down times rather than building additional buildings.

The sustainability of a structure must be balanced with the usefulness of the building, the comfort and safety of its occupants, its aesthetic value, and the economic cost of the building. These are not always in conflict.

There are two common certification programs for sustainable (green) buildings.

Leadership in Energy and Environmental Design (LEED) Certification

The **LEED** certification program was developed by the U.S. Green Building Council (USGBC) and is administered by the Green Building Certification Institute (GBCI). Buildings may be certified as LEED silver, gold, or platinum, with platinum being the highest standard. Certification is based on a points system that looks at the following seven areas of a building's design and construction:

- Sustainable Sites (SS)
- Water Efficiency (WE)
- Energy and Atmosphere (EA)
- Materials and Resources (MR)
- Indoor Environmental Quality (EQ)
- Innovation in Design (ID)
- Regional Prioritization (RP)

For homes, two additional area are added; Location and Linkages (LL) and Awareness and Education (AE).

The GBCI also accredits design and construction professionals. The initial accreditation, the **Green Associate**, provides the professional with a basic understanding of green building and LEED certification. For the advanced accreditation, **LEED AP**, the professional becomes an expert in one of the following areas:

- Building Design and Construction (BD+C)
- Interior Design and Construction (ID+C)
- Home
- Operation and Maintenance (O+M)
- Neighborhood Design (ND)

Visit www.usgbc.org for more information on LEED.

National Green Building Program

The **National Green Building Program** was developed by the National Association of Home Builders (NAHB) and certifies that the home was built in accordance with the International Codes Council (ICC) 700-2008 or the NAHB Model Green Home Building Guidelines. The National Green Building Program seeks to minimize the environmental impact of home construction. Certification is based on the following five areas:

- Energy-efficient features; such as using Energy Star® rated appliances, energy efficient lighting, and renewable sources of energy.
- Water-efficient features, such as low-flow plumbing fixtures.
- Resource-efficient features, such as siting the home to reduce heating, lighting, and cooling costs; and the use of renewable or recycled materials.
- Indoor air quality features, such as heating, ventilation, and air conditioning designed to economically maintain air quality and the use of building material that minimize the release of chemicals, such as volatile organic chemicals (VOCs), into the air.
- Outside the home, such as preserving natural vegetation, landscaping with native plants, and reducing impervious surfaces (such as driveways) that increase water runoff.

For more information on the National Green Building Program visit www.nahbgreen.org.

Building Information Modeling (BIM)

Building information modeling involves building a three-dimensional (3D) computer model of a building before constructing the building in the real world. These models may be used to study the spatial relationship of the building's components and identify any potential conflicts (such as a duct running through a beam) during the design phase where it is easier and cheaper to fix the problem; or they may be used to study another design aspect of a building, such as how the natural lighting changes throughout the day.

Although BIM increases the time and cost of the design phase, these increases are offset by time and cost savings during the construction process through improved coordination and few construction problems resulting from spatial conflicts.

A four-dimensional (4D) model adds the time dimension to a 3D model, allowing design and construction professionals to study and plan how the building will go together during construction. A 4D model can show the anticipated status of a building at any point in the schedule and can identify the order the materials must be delivered to the site for construction.

A five-dimensional (5D) model adds cost to a 4D model, developing a cash flow for the construction project. For the contractor, an accurate cash flow is needed for proper financial management of the project and the construction company. For the owner, an accurate cash flow is needed to obtain financing for the project.

CPM Scheduling (Activity on Node)

In the activity-on-node schedule, the nodes represent the activities or tasks and the arrows represent the relationship between the activities. A typical node is shown below. There is no standard layout for nodes and the reader should verify the node layout when using a schedule. The layout below is a common layout and is the same layout that is used by the American Institute of Constructors (AIC) on the Associate Constructor exam.

ES	NO.	EF
ACTIVITY NAME		
LS	DUR	LF

Definitions:

Activity Name: The name of the activity or task.

Node or Activity Number (No.): The number assigned to the activity. The number may be assigned sequentially (e.g., 1, 2, 3 . . .) or in increments (e.g., 10, 20, 30 . . .).

Duration (DUR): The length of time scheduled for an activity. Construction project are usually scheduled in full days. Shutdowns for maintenance and repairs are usually scheduled in hours or minutes.

Early Start Date (ES): The earliest date an activity can start.

Early Finish Date (EF): The earliest date an activity can finish.

Late Start Date (LS): The latest date an activity can start without delaying the completion of the project.

Late Finish Date (LF): The latest date an activity can finish without delaying the completion of the project.

Total Slack or Float: The amount of time an activity can be delayed without delaying the completion of the project.

Free Slack or float: The amount of time an activity can be delayed without delaying any of the succeeding activities.

Link: The relationship between two activities.

Lag: Used to create delay between activities (positive lag) or overlap activities (negative lag).

Milestone: An activity that marks a key point, such as receiving a notice to proceed. Milestones have a duration of zero.

Drawing a Network Diagram

Most network diagrams are drawn left to right and top to bottom. There is no time scale on a network diagram. Both the personnel who bid the project and the personnel who will manage the project should be involved in its development. A network diagram is drawn as follow:

Step 1: Break the Project Down into Activities. The activities are identified by breaking the project down into large activities, and then breaking the large activities into smaller activities. Activities are broken down until the following criteria are met:

- The activity has a single point of responsibility (i.e., it can be performed by a single trade, crew type, or person).
- The activity has a predictable duration.
- The activity applies to one stage or phase of the project.
- The activity can be performed without interruption.
- Each of the activities must be a manageable size. Activities whose durations are measured in hours are too difficult to track. Activities with a duration measured in months are too broad to be managed.

Step 2: Place the Activities in the Order that they will be Performed. This can be done with Post-it notes so that the activities can be easily moved and rearranged or it may be done in a software package such as Microsoft Project or Primavera P6.

Step 3: Establish Relationships (Links) Between the Activities. Activities are linked with one of the following types of links:

- **Finish-to-Start (FS):** The finish of the predecessor is linked to the start of a succeeding activity. The succeeding activity may start when the preceding activity is completed. This is the most common type of link.

- **Start-to-Start (SS):** The start of the predecessor is linked to the start of a succeeding activity. The succeeding activity may start as soon as the preceding activity has started.

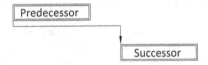

- **Finish-to-Finish (FF):** The finish of the predecessor is linked to the finish of a succeeding activity. The succeeding activity may only finish after the preceding activity is complete.

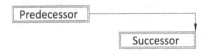

- **Start-to-Finish (SF):** The start of the predecessor is linked to the finish of a succeeding activity. The succeeding activity may finish when the preceding activity has started. This type of link is used to force the succeeding activity to finish just-in-time to start the preceding activity.

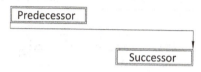

Positive lag is added to the links to create delay between activities and negative lag is added to the links to overlap the activities. For example, a start-to-finish link with two days of positive lag would start the succeeding activity two days after proceeding activity has been completed. Conversely, a two day negative lag would overlap the activity by two days.

All redundant links should be eliminated.

Step 4: Draw the Model. The model is entered into a scheduling software program, if it has not already been done. Alternatively, the model may be drawn by hand. The relationships should be checked to see if they are accurate and make sense.

Example: Draw a network diagram for the following activities:

No.	Name	Predecessors
5	A	
10	B	A
15	C	A
20	D	B, C

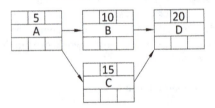

Step 5: Assign Durations to Activities. The estimated labor hours from the project's estimate should be used to estimate the duration of the activities in the schedule. The minimum duration may be estimated by using one of the following equations:

$$\text{Dur} = \frac{\text{Quantity} \times \text{Labor hour per unit}}{\text{Crew size}}$$

$$\text{Dur} = \frac{\text{Quantity}}{\text{Daily crew output}}$$

Example: Determine the minimum required duration for Activity B in the preceding network diagram. Activity B consists of installing 3,300 square feet of block wall. It takes 0.09 labor hours to install one square foot of block wall. The crew consists of three masons and two helpers. The crew works eight hours per day.

$$\text{Dur} = \frac{3,300 \text{ ft}^2 \times 0.09 \text{ lhr/ft}^2}{5 \text{ lhr/hr}} = 59.4 \text{ hr}$$

$$\text{Dur} = \frac{59.4 \text{ hr}}{8 \text{ hr/day}} = 7.4 \text{ Use 8 days}$$

Example: Determine the minimum required duration for Activity C in the preceding network diagram. Activity C consists of framing 1,100 lineal feet of wall. A crew can frame 240 feet of wall per day.

$$\text{Dur} = \frac{1,100 \text{ ft}}{240 \text{ ft/day}} = 4.6 \text{ days use 5 days}$$

The duration may be increased to account for cure time, delays, or other factors.

Example: Update the above network diagram. The duration of Activity A is 2 days, Activity B is 8 days (from previous example), Activity C is 5 days (from previous example), and Activity D is 1 day.

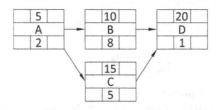

Step 6: Assign Costs and Resources. If resource loading is to be performed or if the schedule is to be used to calculate earned value, cost and resources must be added to the activities.

Step 7: Calculate Start and Finish Dates for Each Activity. For scheduling by hand, the start and finish dates are recorded at the beginning of the day, which is the same as the end of the previous day. Software packages, when scheduling in whole days, schedule the start of the activity at the beginning of a day and the finish at the end of the day.

The early start and finish dates are calculated by a forward pass, which is performed by hand as follows:

- The early start of the first activity is set to 1.
- The early finish is calculated as follows:
$$EF = ES + Dur$$
- The early start for all preceding activity equals the latest (maximum) early finish date of its predecessors.
- Proceed left to right, calculating the early start and early finish dates for all activity.

Example: Determine the early start and early finish dates for the preceding network diagram. Update the network diagram.

$$ES_A = 1$$
$$EF_A = 1 + 2 = 3$$
$$ES_B = 3$$
$$EF_B = 3 + 8 = 11$$
$$ES_C = 3$$
$$EF_C = 3 + 5 = 8$$
$$ES_D = \text{Greater of 11 and 8}$$
$$ES_D = 11$$
$$EF_D = 11 + 1 = 12$$

The network diagram is now as follows:

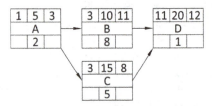

The late start and finish dates are calculated by a backwards pass, which is performed by hand as follows:

- The late finish of the last activity is the same as its early finish date. When there is additional time available to complete the project, the late finish may be set to a later date than the early finish date to use the available time.
- The late start is calculated as follows:

$$LS = LF - Dur$$

- The late finish date for an activity equals the earliest (minimum) late start date of its successors.
- Proceed right to left, calculating the late finish and late start dates for all activities.

Example: Determine the late start and late finish dates for the preceding network diagram. Set the late finish of Activity D equal to its early finish. Update the network diagram.

$$LF_D = 12$$
$$LS_D = 12 - 1 = 11$$
$$LF_C = 11$$
$$LS_C = 11 - 5 = 6$$
$$LF_B = 11$$
$$LS_B = 11 - 8 = 3$$
$$LF_A = \text{Lesser of 3 and 6}$$
$$LF_A = 3$$
$$LS_A = 3 - 2 = 1$$

The network diagram is now as follows:

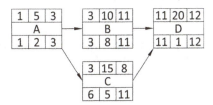

Step 8: Determine Slack (Float) and the Critical Path. The total slack (float) equals the number of days between the early start date and late start date for the activity and is calculated using the following equation:

$$TS = LS - ES$$

Example: Determine the total slack for the activities in the preceding network diagram.

$$TS_A = 1 - 1 = 0$$
$$TS_B = 3 - 3 = 0$$
$$TS_C = 6 - 3 = 3$$
$$TS_D = 11 - 11 = 0$$

The free slack (float) for an activity equals the number of full days between the early finish date for the activity and the earliest early start date of its successors and is calculated using the following equation:

$$FS = \text{Earliest } ES_S - EF$$

Example: Determine the free slack for the activities in the preceding network diagram.

$$FS_A = 3 - 3 = 0$$
$$FS_B = 11 - 11 = 0$$
$$FS_C = 11 - 8 = 3$$

Activity D does not have any predecessors so its free slack is zero.

The critical path includes all activities that have the same total slack (float) as the last activity. If the late finish of the last activity was set to its early finish, it has a total slack of zero and all activities on the critical path will have a total slack of zero.

Example: What is the critical path in the above network diagram?

The critical path is A – B – D.

Step 9: Schedule Activities Start and Finish Times. For activities that have a slack (float), the activities may be scheduled later than their early start and early finish dates, using some or all of the slack (float), to level resources. In the above network diagram, Activities A, B, and D do not have any slack (float), so they must start on schedule or the completion date will slip. Activity C may start anytime between the beginning of the 3rd and 6th day, without the schedule slipping.

Safety

Construction safety is regulated by state or federal Occupational Health and Safety Administration (OSHA). Some earthwork operations, such as gravel pits, are regulated by the Mine Safety and Health Administration (MSHA). Additional safety provisions may be required by the owner or the general contractor. The federal OSHA regulations are found in the Code of Federal Regulations Title 29, Part 1926 and can be downloaded at www.gpoaccess.gov/cfr/index.html.

The contractor is responsible for safety on the jobsite including establishing procedures for analyzing and preventing (or controlling if not preventable) jobsite hazards, providing the necessary safety equipment, and providing safety training to workers. At a minimum, the contractor's safety plan should address the following:

The minimum safety **Training** required before a worker can work on the jobsite. The format and attendance requirements for **Safety Meetings** and ongoing training.

The procedures for maintaining and making Material Safety Data Sheets (**MSDS**) available to jobsite workers.

The minimum **Personal Protective Equipment (PPE)** required for each type of job, including eye and face protection, hearing protection, head protection, foot protection, and personal fall protection.

Cleanliness and **Housekeeping** at the jobsite, including daily cleanup and fire prevention.

The minimum design and inspection requirement for temporary **Stairs and Ladders**, including the use of guardrails.

The minimum design and inspection requirements for **Scaffolding and Shoring**, including proper planking, toe boards, and the use of guardrails.

Fall Protection, such as guard rails, around openings in the floor, windows openings, and the edge of the roof.

The safety procedures for working in **Trench** and **Foundation Excavation**, including the use of trench boxes in narrow trenches.

The safety and inspection requirements for **Hand Tools**, including how to remove a tool from service.

The safety requirements for working in and around **Vehicles** and **Mobile Equipment**, including the use of backup alarms and rollover protection.

Electrical safety, including **Lockout/Tagout** procedures, the use of extension cords, and that all circuits are protected by ground fault circuit interrupters (GFCIs).

The employees are responsible for following all of the safety rules, wearing the required personal protective equipment, avoiding unsafe practices (such as horse play), placing unsafe tools (such as dull tools or tools missing safety guards) out of service, and reporting any unsafe practice or condition to management.

Present Value, Future Value, and Annual Equivalent

The calculation of present value, future value, and annual equivalent is based upon the concept of equivalence. Equivalent cash flows are cash flows that produce the same result over a specific period of time and are based upon (1) the size of the cash flows, (2) the timing of the cash flows, and (3) a specific interest rate. The equations use the following variables.

P = Present value or value of the cash flow at the present time

F = Future value or value of the cash flow at the some future time

i = The periodic interest rate or the interest rate for one period, expressed in decimal format (e.g., 1% is 0.01)

n = The number of interest compounding periods, usually months or years

A = The value of a single cash flow in a uniform series of cash flows that occurs at the end of each period and continues for n number of periods

APR = Annual percentage rate

Monthly Interest Rate

The monthly interest rate is calculated using the following equation:

$$i = \frac{\text{APR}}{12}$$

Example: Determine monthly interest rate for a bank account with an annual percentage rate of 6%.

$$i = \frac{6\%}{12} = \frac{0.06}{12} = 0.005$$

Single-Payment Compound-Amount Factor

Converts a present value into a future value using the following equation:

$$F = P(1+i)^n$$

Example: Determine the future value at the end of five years (60 months) of $100 deposited in a bank account which earns 0.5% interest per month.

$$F = \$100(1+0.005)^{60} = \$134.89$$

Single-Payment Present-Worth Factor

Converts a future value into a present value using the following equation:

$$P = \frac{F}{(1+i)^n}$$

Example: How much would you need to deposit in a bank account today to have $500 ten years (120 months) from now? The bank account earns 0.4% interest per monthly.

$$P = \frac{\$500}{(1+0.004)^{120}} = \$309.69$$

Uniform-Series Compound-Amount Factor

Converts a uniform series into a future value using the following equation:

$$F = A\left[\frac{(1+i)^n - 1}{i}\right]$$

Example: $100 is deposited at the end of each month in a bank account for the next five years (60 months). What it the value of the deposits at the end of five years if the bank account earns 0.45% per month?

$$F = \$100\left[\frac{(1+0.0045)^{60} - 1}{0.0045}\right] = \$6,870.47$$

Uniform-Series Sinking-Fund Factor

Converts a future value into a uniform series using the following equation:

$$A = F\left[\frac{i}{(1+i)^n - 1}\right]$$

Example: At the end of each month for the next 40 years (480 months) you make a deposit in a retirement account that earns 0.55% interest per month. How much do you need to deposit each month to have $1,000,000 at the end of the 40 years?

$$A = \$1,000,000\left[\frac{0.0055}{(1+0.0055)^{480} - 1}\right] = \$425.95$$

Uniform-Series Present-Worth Factor

Converts a uniform series into a present worth using the following equation:

$$P = A\left[\frac{(1+i)^n - 1}{i(1+i)^n}\right]$$

Example: How much would you need to deposit in a bank account today to be able to withdraw $250 per month for the next 20 years (240 months)? The bank account earns 0.4% interest per month.

$$P = \$250\left[\frac{(1+0.004)^{240} - 1}{0.004(1+0.004)^{240}}\right] = \$38,523.33$$

Uniform-Series Capital-Recovery Factor

Converts a present worth into a uniform series using the following equation:

$$A = P\left[\frac{i(1+i)^n}{(1+i)^n - 1}\right]$$

Example: Determine the monthly payment on a 30-year (360-month) loan with an interest rate of 0.6% per month and a principal of $200,000.

$$A = \$200,000\left[\frac{0.006(1+0.006)^{360}}{(1+0.006)^{360} - 1}\right] = \$1,357.58$$

Complex Cash Flows

Cash flows that occur in the same period of time may be added and subtracted. Complex cash flows that cannot be handled by one of the above equations may be handled by converting the different components of the cash flow to a common time period and adding or subtracting the cash flows.

Example: Determine the present value of $100 paid one year from now and $200 paid two years from now using an annual interest rate of 6%.

$$P_1 = \frac{\$100}{(1+0.06)^1} = \$94.34$$

$$P_2 = \frac{\$200}{(1+0.06)^2} = \$178.00$$

$$P = P_1 + P_2 = \$94.34 + \$178.00 = \$272.34$$

Life Cycle Cost Analysis

The life cycle cost is the total cost of a piece of equipment or building over its entire life. The cost includes its initial costs (e.g., the purchase of the equipment, its installation, startup, etc.), its operational costs (e.g., repairs, maintenance, energy, etc.), and its decommissioning costs (e.g., removal and disposal). The life cycle cost is often used to select between two or more alternatives. The life cycle cost for an alternative is calculated by determining the present value of all of the alternative's costs. The alternative with the lowest life cycle cost is chosen. For a comparison to be made, the useful life of each option must be the same or the comparison must be made based upon the annual equivalent.

Example: Your company needs to purchase a new furnace and has narrowed the choices down to two furnaces. The first furnace costs $3,000 and has an estimated annual operating cost of $1,050. The second furnace costs $3,500 and has an estimated annual operating cost of $950. The estimated life of both furnaces is fifteen years. Using an interest rate of 15% per year, which furnace has the cheapest life cycle cost?

$$P_1 = -\$3,000 - \$1,050 \left[\frac{(1+0.15)^{15} - 1}{0.15(1+0.15)^{15}} \right] = -\$9,140$$

$$P_2 = -\$3,500 - \$950 \left[\frac{(1+0.15)^{15} - 1}{0.15(1+0.15)^{15}} \right] = -\$9,055$$

The second furnace has a cheaper life cycle cost.

Earned Value

Earned value is a way to measure the schedule and cost performance of a project. Earned value uses the following variables:

BCWS = Budgeted cost of work scheduled. The BCWS is calculated by multiplying the percent each activity is supposed to be complete according to the schedule by its budget, then summing the costs.

BCWP = Budgeted cost of work performed. The BCWP is calculated by multiplying the actual percent complete for each activity by its budget, then summing the cost.

ACWP = Actual cost of work performed. The ACWP is obtained from the accounting system.

Schedule Performance Index (SPI)

The SPI is used to measure how the project's actual schedule compares to the planned schedule, including all activities with a cost in the measure. One of the weaknesses is that the SPI excludes all activities that have a cost of zero. The SPI is calculated as follows:

$$SPI = \frac{BCWP}{BCWS}$$

An SPI greater than 1 indicates that the project is ahead of schedule. An SPI equal to 1 indicates that the project is on schedule. And an SPI less than 1 indicates the project is behind schedule.

Cost Performance Index (CPI)

The CPI is used to measure how the project's actual costs compare to the budget. Care must be taken to make sure that all of the actual costs have been entered into the accounting system. Not having recorded all of the costs associated with the work will render the CPI useless. The CPI is calculated as follows:

$$CPI = \frac{BCWP}{ACWP}$$

A CPI greater than 1 indicates that the project is under budget. A CPI equal to 1 indicates that the project is on budget. And a CPI less than 1 indicates the project is over budget.

Example: A project consists of the following three activities. The start and finish dates are at the beginning of the day. At the beginning of day 8, Activity A is 75% complete, Activity B is 22% complete, and Activity C is 38% complete. The actual cost of the work performed is $1,810. Determine the CPI and SPI.

Activity	Start	Finish	Cost
A	Day 1	Day 11	$1,000
B	Day 6	Day 16	$1,500
C	Day 6	Day 11	$2,000

The percent each activity was supposed to be complete based upon the schedule is calculated as follows.

$$\text{Activity A} = \frac{(8-1)}{(11-1)} = 0.7 \text{ or } 70\%$$

$$\text{Activity B} = \frac{(8-6)}{(16-6)} = 0.2 \text{ or } 20\%$$

$$\text{Activity C} = \frac{(8-6)}{(11-6)} = 0.4 \text{ or } 40\%$$

The BCWS is calculated as follows.

$$\text{BCWS} = 0.7 \times \$1,000 + 0.2 \times \$1,500 + 0.4 \times \$2,000$$
$$\text{BCWS} = \$1,800$$

The BCWP is calculated as follows.

$$\text{BCWP} = 0.75 \times \$1,000 + 0.22 \times \$1,500 + 0.38 \times \$2,000$$
$$\text{BCWP} = \$1,840$$

The SPI is calculated as follows:

$$\text{SPI} = \frac{\$1,840}{\$1,800} = 1.022$$

The project is ahead of schedule. The CPI is calculated as follows.

$$\text{CPI} = \frac{\$1,840}{\$1,810} = 1.016$$

The project is under budget.

Chapter 8: Ethics

Ethics can be defined by two broad definitions: (1) a set of values or a guiding philosophy of the individual and (2) a set of principles or rules that members of a group agree to abide by. Each person must evaluate and define her own guiding philosophy. When organizations establish a set of rules, they often publish them for their members and the public to see. The following are two sets of ethical principles that are specific to the construction industry.

Code of Professional Ethics for the Construction Manager

Since 1982, the Construction Management Association of America (CMAA) has taken a leadership role in regard to critical issues impacting the construction and program management industry, including the setting of ethical standards of practice for the Professional Construction Manager.

The Board of Directors of CMAA has adopted the following Code of Professional Ethics of the Construction Manager (CODE) which apply to CMAA members in performance of their services as Construction and Program Managers. This Code applies to the individuals and to organizations who are members of CMAA.

All members of the Construction Management Association of America commit to conduct themselves and their practice of Construction and Program Management in accordance with the Code of Professional Ethics of the Construction Manager.

As a professional engaged in the business of providing construction and program management services, and as a member of CMAA, I agree to conduct myself and my business in accordance with the following:

1. **Client Service.** I will serve my clients with honesty, integrity, candor, and objectivity. I will provide my services with competence, using reasonable care, skill and diligence consistent with the interests of my client and the applicable standard of care.

2. **Representation of Qualifications and Availability.** I will only accept assignments for which I am qualified by my education, training, professional experience and technical competence, and I will assign staff to projects in accordance with their qualifications and commensurate with the services to be provided, and I will only make representations concerning my qualifications and availability which are truthful and accurate.

3. **Standards of Practice.** I will furnish my services in a manner consistent with the established and accepted standards of the profession and with the laws and regulations which govern its practice.

4. **Fair Competition.** I will represent my project experience accurately to my prospective clients and offer services and staff that I am capable of delivering. I will develop my professional reputation on the basis of my direct experience and service provided, and I will only engage in fair competition for assignments.

5. **Conflicts of Interest.** I will endeavor to avoid conflicts of interest; and will disclose conflicts which in my opinion may impair my objectivity or integrity.

6. **Fair Compensation.** I will negotiate fairly and openly with my clients in establishing a basis for compensation, and I will charge fees and expenses that are reasonable and commensurate with the services to be provided and the responsibilities and risks to be assumed.

7. **Release of Information.** I will only make statements that are truthful, and I will keep information and records confidential when appropriate and protect the proprietary interests of my clients and professional colleagues.

8. **Public Welfare.** I will not discriminate in the performance of my Services on the basis of race, religion, national origin, age, disability, or sexual orientation. I will not knowingly violate any law, statute, or regulation in the performance of my professional services.

9. **Professional Development.** I will continue to develop my professional knowledge and competency as Construction Manager, and I will contribute to the advancement of the construction and program management practice as a profession by fostering research and education and through the encouragement of fellow practitioners.

10. **Integrity of the Profession.** I will avoid actions which promote my own self-interest at the expense of the profession, and I will uphold the standards of the

construction management profession with honor and dignity.

Courtesy of Construction Management Association of America

American Society of Professional Estimators (ASPE) Code of Ethics – Basic Canons

Canon #1 - Professional estimators shall perform services in areas of their discipline and competence.

Canon #2 - Professional estimators shall continue to expand their professional capabilities through continuing education programs to better enable them to serve clients, employers and the industry.

Canon #3 - Professional estimators shall conduct themselves in a manner, which will promote cooperation and good relations among members of our profession and those directly related to our profession.

Canon #4 - Professional estimators shall safeguard and keep in confidence all knowledge of the business affairs and technical procedures of an employer or client.

Canon #5 - Professional estimators shall conduct themselves with integrity at all times and not knowingly or willingly enter into agreements that violate the laws of the United States of America or of the states in which they practice. They shall establish guidelines for setting forth prices and receiving quotations that are fair and equitable to all parties.

Canon #6 - Professional estimators shall utilize their education, years of experience and acquired skills in the preparation of each estimate or assignment with full commitment to make each estimate or assignment as detailed and accurate as their talents and abilities allow.

Canon #7 - Professional estimators shall not engage in the practice of "bid peddling" as defined by this code. This is a breach of moral and ethical standards, and a member of this society shall not enter into this practice.

Canon #8 - Professional estimators and those in training to be estimators shall not enter into any agreement that may be considered acts of collusion or conspiracy (bid rigging) with the implied or express purpose of defrauding clients. Acts of this type are in direct violation of the Code of Ethics of the American Society of Professional Estimators.

Canon #9 - Professional estimators and those in training to be estimators shall not participate in acts, such as the giving or receiving of gifts, that are intended to be or may be construed as being unlawful acts of bribery.

Courtesy of the American Society of Professional Estimators

Chapter 9: Construction Materials

Concrete
Cement Types

Type I - Normal cement
Type II - Moderate-sulfate-resistance cement
Type III - High-early-strength cement
Type IV - Low-heat-of-hydration cement
Type V - High-sulfate-resistance cement

Rebar Properties

Bar Size	Nominal Diameter (in)	Cross-sectional Area (in^2)	Weight (pound per ft)
2	1/4"	0.05	0.167
3	3/8"	0.11	0.376
4	1/2"	0.20	0.668
5	5/8"	0.31	1.043
6	3/4"	0.44	1.502
7	7/8"	0.60	2.044
8	1"	0.79	2.670
9	1-1/8"	1.00	3.400
10	1-1/4"	1.27	4.303
11	1-3/8"	1.56	5.313

Rebar Lap Percentage

	Percentage to Add for Rebar Lap		
		Bar Length	
Lap	20 ft	40 ft	60 ft
12 in	6%	3%	2%
15 in	7%	4%	3%
18 in	9%	4%	3%
21 in	10%	5%	4%
24 in	12%	6%	4%
27 in	13%	6%	4%
30 in	15%	7%	5%
33 in	16%	8%	5%
36 in	18%	9%	6%

Welded Wire Fabric (WWF)

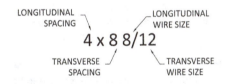

LONGITUDINAL SPACING — LONGITUDINAL WIRE SIZE

$$4 \times 8 \; 8/12$$

TRANSVERSE SPACING — TRANSVERSE WIRE SIZE

Percentage to Add for WWF Lap		
Lap	8' x 20' Sheet	5' x 150' Roll
6 in	10%	12%
12 in	21%	26%

Foundation Walls and Continuous Footings

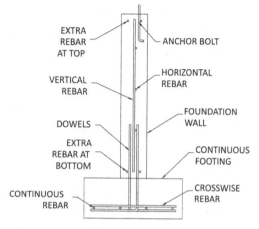

A **Footing** is used to disperse the building's loads (forces created by the building) to the surrounding soil, which has a lower capacity to support these loads than concrete and other building materials. Concrete has a high compressive strength, but is weak in tension. Rebar is added to concrete to resist tensile forces (tension).

A **Continuous Footing** is used to support foundation walls. Continuous footings often include **Continuous Rebar**, which runs the length of the footing, **Crosswise Rebar**, which runs the width of the footing, and **Dowels**, which connect the continuous footing to the foundation wall.

The quantity of concrete needed is estimated by determining the volume of the footing.

The lineal feet of **Continuous Rebar** is estimated by multiplying the number of bars by the length of the footing and then adding the rebar lap percentage from the preceding Rebar Lap Percentage table.

The number of pieces of **Crosswise Rebar** needed is estimated by dividing the length of the continuous footing by the spacing between the bars and adding one bar for each wall segment. The additional bar per wall segment allows for an extra bar for each corner. The length of the crosswise rebar is equal to the width of the footing less twice the required cover (typically 3 inches per side).

The number of **Dowels** needed is estimated by dividing the length of the foundation wall by the spacing between the bars and adding one dowel for each wall segment.

The **Foundation Wall** often includes **Vertical Rebar** running up and down in the wall and **Horizontal Rebar**, which runs horizontally along the length of the wall. **Extra Top and Bottom Bars** are often added at the top and bottom of the wall. **Anchor Bolts** are embedded in the top of the wall to connect the wall to framing or bearing plates. Embedded steel plates (**Embeds**) are often used to connect the structure to the foundation wall.

The number of **Vertical Rebar** needed is estimated by dividing the length of the foundation wall by the spacing between the bars and adding one bar for each wall segment. The length of the vertical rebar will be 2 to 3 inches shorter than the height of the wall to ensure that the bar will not extend out of the wall.

The number of **Horizontal Rebar** needed is estimated by dividing the wall height by the spacing between the bars and adding one as a starting bar at the bottom of the wall. If **Extra Top and Bottom Bars** are required they are added to the number of horizontal bars. If two bars are needed at the top, one bar is added because one bar is already included in the number of horizontal bars. The total length of horizontal bars, including the extra top and bottom bars, is estimated by multiplying the number of bars by the length of the wall and then adding the rebar lap percentage from the preceding Rebar Lap Percentage table.

The number of **Anchor Bolts** needed is estimated by dividing the length of the foundation wall by the spacing between the bolts and adding one bolt for each wall segment. The number of **Embeds** is estimated by counting them.

Example: Determine the materials needed for 100 feet of 4-foot-high by 8-inch-thick concrete wall that sits on a 24-inch-wide by 12-inch-high footing. There are four wall segments. The footing is reinforced with four #4 continuous bars, #4 crosswise bars at 18 inches on center, and #5 dowels at 18 inches on center. The wall is reinforced with #4 continuous rebar running horizontally at 12 inches on center and #4 rebar running vertically at 18 inches on center. Two #4 rebar are required at the top and bottom of the wall. Anchor bolts, spaced 24 inches on center, are used to connect the concrete wall to the wood framing. The continuous rebar is ordered in 20 foot lengths and is lapped 18 inches.

The following materials are required for the footing:

$$\text{Concrete} = (100 \text{ ft})\left(\frac{12 \text{ in}}{12 \text{ in/ft}}\right)\left(\frac{24 \text{ in}}{12 \text{ in/ft}}\right)\left(\frac{1 \text{ yd}^3}{27 \text{ ft}^3}\right)$$

$$\text{Concrete} = 7.4 \text{ yd}^3$$

$$\text{Rebar}_{\text{Crosswise}} = \frac{(100 \text{ ft})(12 \text{ in/ft})}{18 \text{ in}} + 4 = 71 \text{ ea}$$

$$\text{Rebar}_{\text{Continuous}} = (4 \text{ each})(100 \text{ ft})\left(1 + \frac{9}{100}\right) = 436 \text{ ft}$$

$$\text{Dowels} = \frac{(100 \text{ ft})(12 \text{ in/ft})}{18 \text{ in}} + 4 = 71 \text{ ea}$$

The crosswise bars are 18 inches (24 in − 2 × 3 in) long. The following materials are required for the foundation wall:

$$\text{Concrete} = (100 \text{ ft})(4 \text{ ft})\left(\frac{8 \text{ in}}{12 \text{ in/ft}}\right)\left(\frac{1 \text{ yd}^3}{27 \text{ ft}^3}\right)$$

$$\text{Concrete} = 9.9 \text{ yd}^3$$

$$\text{Number}_{\text{Horizontal}} = \frac{(4 \text{ ft})}{(1 \text{ ft})} + 1 + 2 = 7 \text{ ea}$$

$$\text{Rebar}_{\text{Horizontal}} = (7 \text{ ea})(100 \text{ ft/ea})\left(1 + \frac{9}{100}\right) = 763 \text{ ft}$$

$$\text{Rebar}_{\text{Vertical}} = \frac{(100 \text{ ft})(12 \text{ in/ft})}{18 \text{ in}} + 4 = 71 \text{ ea}$$

$$\text{Anchor bolts} = \frac{(100 \text{ ft})(12 \text{ in/ft})}{24 \text{ in}} + 4 = 54 \text{ ea}$$

The vertical bars are 3 foot 10 inches long (4 ft − 2 in).

Spread Footing and Column

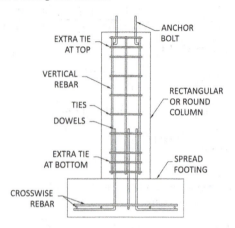

A **Spread Footing** is used to support a column or short foundation wall. **Dowels** are used to connect the footing to concrete columns and walls. Spread footings may include **Crosswise Rebar** running in one or both directions.

The number of pieces of **Crosswise Rebar** needed is estimated by counting the number of bars required per footing and multiplying it by the number of footings. Be sure to count bars running in both directions. The length of the crosswise rebar is equal to the width of the footing less twice the required cover (typically 3 inches per side).

The number of **Dowels** needed is estimated by counting the number of dowels required per column and multiplying it by the number of columns.

The **Column** includes **Vertical Bars** running the height of the column and **Ties** to keep the bars from spreading.

The number of **Vertical Rebar** needed is estimated by counting the number of bars required per column and multiplying it by the number of columns. The length of the vertical rebar will be 2 to 3 inches shorter than the height of the column to ensure that the bar will not extend out of the column.

The number of **Ties** needed is estimated by dividing the column height by the spacing between the ties and adding one as a starting tie at the bottom of the column. If **Extra Top and Bottom Ties** are required they are added to the number of ties. The total number of ties is estimated by multiplying the number of ties per column by the number of columns. The size of the ties is equal to the width or diameter of the column less twice the required cover (typically 2 inches per side).

The footing or column is connected to steel columns by **Anchor Bolts** and to wood columns by **Framing Hardware**, such as a post base.

The number of **Anchor Bolts** needed is estimated by counting the number of anchor bolts required per column and multiplying it by the number of columns. **Framing Hardware** is estimated by counting each type of hardware.

Example: A project requires eight 12-inch by 12-inch by 6-foot-high columns, which are supported by 30-inch by 30-inch by 12-inch-high spread footings. The footings are reinforced with five #4 crosswise bars in both directions. Four #4 dowels are used to connect the footing to the column. The columns are reinforced with four #6 vertical bars and #3 ties at 12 inches on center. An additional tie is required at the top and bottom. Four anchor bolts are used to connect the concrete column to a steel column.

The following materials are required for the footings:

$$\text{Concrete} = (8 \text{ ea})\left(\frac{30 \text{ in}}{12 \text{ in/ft}}\right)\left(\frac{30 \text{ in}}{12 \text{ in/ft}}\right)\left(\frac{12 \text{ in}}{12 \text{ in/ft}}\right)\left(\frac{1 \text{ yd}^3}{27 \text{ ft}^3}\right)$$

$$\text{Concrete} = 1.9 \text{ yd}^3$$

$$\text{Rebar}_{\text{Crosswise}} = (8 \text{ ea})(5 \times 2 \text{ per ea}) = 80 \text{ ea}$$

$$\text{Dowels} = (8 \text{ ea})(4 \text{ per ea}) = 32 \text{ ea}$$

The crosswise bars are 24 inches (30 in − 2 × 3 in) long. The following materials are required for the columns:

$$\text{Concrete} = (8 \text{ ea})\left(\frac{12 \text{ in}}{12 \text{ in/ft}}\right)\left(\frac{12 \text{ in}}{12 \text{ in/ft}}\right)(6 \text{ ft})\left(\frac{1 \text{ yd}^3}{27 \text{ ft}^3}\right)$$

$$\text{Concrete} = 1.8 \text{ yd}^3$$

$$\text{Ties} = (8 \text{ ea})\left(\frac{(6 \text{ ft})(12 \text{ in/ft})}{(12 \text{ in})} + 1 + 2\right) = 72 \text{ ea}$$

$$\text{Rebar}_{\text{Vertical}} = (8 \text{ ea})(4 \text{ per ea}) = 32 \text{ ea}$$

$$\text{Anchor bolts} = (8 \text{ ea})(4 \text{ per ea}) = 32 \text{ ea}$$

The vertical bars are 5 foot 10 inches long (6 ft − 2 in) and the ties are 8 inches (12 in − 2 × 2 in) square.

Slab on Grade

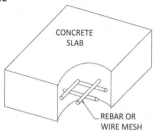

Concrete slabs consist of the **Concrete Slab**, which may be reinforced with **Rebar** or **Welded Wire Fabric**, and may be poured over a vapor barrier or layer of gravel. The slab may be thickened at entrances and in other areas to act as a footing.

The quantity of concrete needed is estimated by multiplying the area of the slab by its thickness. The thickened portions of the slab are estimated by multiplying their cross-sectional areas by their lengths.

The lineal feet of **Rebar** is estimated by dividing the slab into rectangular areas. For a rectangular area, the number of runs of rebar in the long (or short) direction is estimated by dividing the width (or length) by the spacing between the bars and adding one as a starting bar. The lineal feet of rebar in the long (or short) direction is estimated by multiplying the number of runs by the length (or width) of the slab. The total length of rebar is estimated by adding the rebar in the long and short directions together and then adding the rebar lap percentage from the preceding Rebar Lap Percentage table.

Example: Determine the materials needed for a 60-foot-long by 40-foot-wide by 4-inch-thick concrete slab. The slab is reinforced with #4 rebar spaced at 18 inches on center. The rebar is ordered in 20 foot lengths and is lapped 18 inches.

$$\text{Area} = (60 \text{ ft})(40 \text{ ft}) = 2,400 \text{ ft}^2$$

$$\text{Concrete} = \left(2,400 \text{ ft}^2\right)\left(\frac{4 \text{ in}}{12 \text{ in/ft}}\right)\left(\frac{1 \text{ yd}^3}{27 \text{ ft}^3}\right) = 29.6 \text{ yd}^3$$

$$\text{Runs}_{\text{Length}} = \frac{(60 \text{ ft})(12 \text{ in/ft})}{(18 \text{ in})} + 1 = 41 \text{ runs}$$

$$\text{Rebar}_{\text{Length}} = (41 \text{ runs})(40 \text{ ft/run}) = 1,640 \text{ ft}$$

$$\text{Runs}_{\text{Width}} = \frac{(40 \text{ ft})(12 \text{ in/ft})}{(18 \text{ in})} + 1 = 28 \text{ runs}$$

$$\text{Rebar}_{\text{Width}} = (28 \text{ runs})(60 \text{ ft/run}) = 1,680 \text{ ft}$$

$$\text{Rebar} = (1,640 \text{ ft} + 1,680 \text{ ft})\left(1 + \frac{9}{100}\right) = 3,619 \text{ ft}$$

$$\text{Rebar} = \frac{3,619 \text{ ft}}{20 \text{ ft/bar}} = 181 \text{ bars}$$

The quantity of **Welded Wire Fabric** can be estimated by adding the lap factor to the area of the slab.

Example: How much welded wire fabric is needed for an 80-foot-long by 35-foot-wide concrete slab? The mesh is ordered in 5-foot by 150-foot rolls and is lapped 6 inches.

$$\text{Area} = (80 \text{ ft})(35 \text{ ft}) = 2{,}800 \text{ ft}^2$$

$$\text{WWF} = (2{,}800 \text{ ft}^2)\left(1 + \frac{12}{100}\right) = 3{,}136 \text{ ft}^2$$

$$\text{WWF} = \frac{3{,}136 \text{ ft}^2}{(5 \text{ ft} \times 150 \text{ ft/roll})} = 4.1 \text{ rolls} \quad \text{Order 5 rolls}$$

Forming

Formwork must be designed for its specific applications and is beyond the scope of this book.

Formwork for footings and slabs are often bid by the lineal foot. Care must be taken to include all sides and ends.

Example: Determine the length of the forms needed for 100 feet of 24-inch-wide by 12-inch-high footing.

$$\text{Length} = (2 \text{ side})(100 \text{ ft/side}) + (2 \text{ ends})(2 \text{ ft/end})$$

$$\text{Length} = 204 \text{ ft}$$

Often formwork is bid by the square foot contact area (SFCA); the area that comes in contact with the form. When bidding this way, one must be sure to include all surfaces (sides, ends, and if the concrete is not in contact with the ground, the bottom) and use standard form dimensions. If a wall is 42 inches high and the forms come in two foot increments, the height of the forms will be 4 feet, and 4 feet should be used rather than the actual height of the wall.

Example: Determine the area of the forms needed for 100 feet of 4-foot-high by 8-inch-thick concrete wall.

$$\text{SFCA} = (2 \text{ sides})(100 \text{ ft/side})(4 \text{ ft})$$

$$+ (2 \text{ ends})\frac{(8 \text{ in/ends})}{(12 \text{ in/ft})}(4 \text{ ft})$$

$$\text{SFCA} = 806 \text{ ft}^2$$

Masonry
Concrete Masonry Unit (CMU)

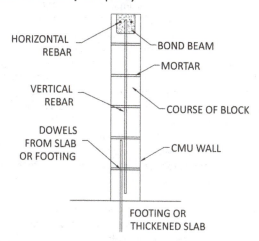

HORIZONTAL REBAR

BOND BEAM

MORTAR

VERTICAL REBAR

COURSE OF BLOCK

DOWELS FROM SLAB OR FOOTING

CMU WALL

FOOTING OR THICKENED SLAB

CMU walls consist of concrete blocks laid in courses (rows). The blocks are 15 5/8 inches long by 7 5/8 inches high and are placed with 3/8-inch mortar joints, which make the block and mortar joints 16 inches long by 8 inches high. The blocks are available in a number of widths, including 3 5/8, 5 5/8, 7 5/8, 9 5/8, and 11 5/8 inches. A few of the specialty available blocks are shown below.

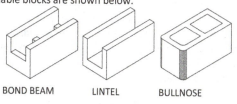

BOND BEAM   LINTEL   BULLNOSE

The blocks are reinforced by **Vertical Rebar** placed in the cells of the blocks and **Horizontal Rebar** placed in **Bond Beams.** Alternately, the horizontal reinforcement may be provided by wire ladders placed in the mortar joints. The vertical rebar must be short enough that masons can lift the block over the top of the bar. The typical length for the vertical rebar is equal to the spacing between the bond beams plus the required lap for the rebar, except for the top row of bars whose length is equal to the distance between the last two bond beams less a couple of inches so that the bar does not stick out of the top of the wall. The bond beams consist of horizontal rebar grouted in bond beam blocks. CMU walls are connected to the underlying footing or thickened slab using dowels that are poured into the footing or slab.

When estimating the number of blocks, the bond beam and any other specialty blocks must be kept separate from standard blocks. The number of blocks per row is determined by dividing the length of the wall by length of the block and a mortar joint (16 inches). The number of rows is determined by dividing the height of the wall by the height of a block and a mortar joint (8 inches). Both of these numbers should be rounded up to a whole number. The number of blocks required is the number of rows multiplied by the number of blocks per row. Rebar is estimated as it was in a concrete wall, except you do not add one when figuring the horizontal rebar. Mortar and grout are estimated based upon the number of cubic feet needed for 100 square feet of block, which is obtained from historical data.

Example: Determine the number of blocks needed for 50 feet of 8-foot-high CMU wall. The wall is reinforced vertically with #5 rebar at 32 inches on center and horizontally with two #4 rebar at 48 inches on center. The horizontal rebar is ordered in 20 foot lengths. All rebar is lapped 18 inches. Nine cubic feet of mortar and 18 cubic feet of grout are required for 100 square feet of wall.

$$\text{Blocks per row} = \frac{(50 \text{ ft})(12 \text{ in/ft})}{(16 \text{ in})} = 38 \text{ blocks/row}$$

$$\text{Rows} = \frac{(8 \text{ ft})(12 \text{ in/ft})}{(8 \text{ in})} = 12 \text{ rows}$$

Two of these rows, one at 4 foot and one at 8 foot, will require bond beam blocks.

$$\text{Blocks}_{\text{Standard}} = (38 \text{ blocks/row})(10 \text{ rows}) = 380 \text{ blocks}$$

$$\text{Blocks}_{\text{Bond beam}} = (38 \text{ blocks/row})(2 \text{ rows}) = 76 \text{ blocks}$$

$$\text{Rebar}_{\text{Horizontal}} = (2/\text{bond beam})(2 \text{ bond beams}) = 4 \text{ ea}$$

$$\text{Rebar}_{\text{Horizontal}} = (4 \text{ ea})(50 \text{ ft})\left(1 + \frac{9}{100}\right) = 218 \text{ ft}$$

$$\text{Number}_{\text{Vertical Rebar}} = \frac{(50 \text{ ft})(12 \text{ in/ft})}{(32 \text{ in})} + 1 = 20 \text{ ea}$$

Order 20 #5 rebar, 66 inches long (48 in + 18 in) and 20 #5, 46 inches long (48 in − 2 in) for the vertical rebar.

$$\text{Mortar} = (50 \text{ ft})(8 \text{ ft})(9 \text{ ft}^3/100 \text{ ft}^2) = 36 \text{ ft}^3$$

$$\text{Grout} = (50 \text{ ft})(8 \text{ ft})(18 \text{ ft}^3/100 \text{ ft}^2) = 72 \text{ ft}^3$$

Brick Terminology

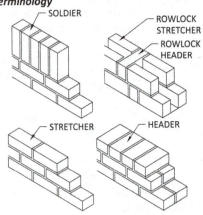

SOLDIER

ROWLOCK STRETCHER

ROWLOCK HEADER

STRETCHER

HEADER

Mortar Joints

CONCAVE

RAKED

STRUCK

"V"

FLUSH

WEATHERED

Brick Bonds (Patterns)

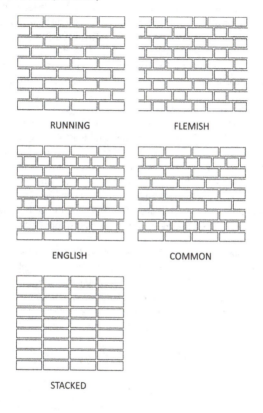

RUNNING

FLEMISH

ENGLISH

COMMON

STACKED

Brick Veneer

Brick Veneer consists of a layer of face brick installed over a wood or block wall and is secured to the wall using metal ties. Wood walls must be covered with a vapor barrier (e.g., Tyvek). Weep holes are placed at the base of the brick veneer to allow the moisture that gets between the walls to drain out. Face brick comes in a variety of sizes.

The number of bricks (units) required may be estimated the same way that the number of blocks was estimated. Alternately, the number of bricks can be estimated by multiplying the area of the brick veneer by the number of bricks per square foot. Mortar is estimated in the same way as it was for blocks. The number of bricks and quantity of mortar per square foot often can be obtained from the brick manufacturer. The specifications or building code set the requirement for the number of brick ties.

Example: Determine the number of bricks needed for 20 feet of 3-foot-high brick veneer. A square foot of brick veneer requires 6.75 bricks and 0.3 cubic feet of mortar. One tie is required for every 2 square feet of brick. Add 5% waste to the brick and 10% waste to the mortar and ties.

$$\text{Area} = (3 \text{ ft})(20 \text{ ft}) = 60 \text{ ft}^2$$

$$\text{Bricks} = (60 \text{ ft}^2)(6.75 \text{ ea/ft}^2)\left(1 + \frac{5}{100}\right) = 425 \text{ ea}$$

$$\text{Mortar} = (60 \text{ ft}^2)(0.3 \text{ ft}^3/\text{ft}^2)\left(1 + \frac{10}{100}\right) = 20 \text{ ft}^3$$

$$\text{Ties} = \frac{(60 \text{ ft}^2)}{(2 \text{ ft}^2/\text{tie})}\left(1 + \frac{10}{100}\right) = 33 \text{ ea}$$

Steel
Commons Structural Steel Shapes

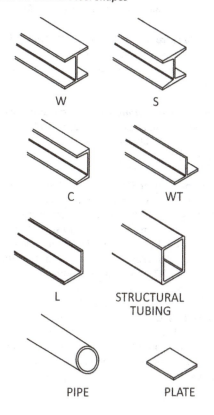

W

S

C

WT

L

STRUCTURAL
TUBING

PIPE

PLATE

Wide-Flange Beams (W) are used as columns, girders, and beams. A wide-flange beam is designated by a "W" followed by its nominal depth and weight per foot.

American Standard Beams (S) are sometimes used as girders and beams. An American standard beam is designated by an "S" followed by its nominal depth and weight per foot.

American Standard Channels (C) are used as beams and stair stringers. An American standard channel is designated by a "C" followed by its nominal depth and weight per foot.

Structural Tees (WT or ST) are used in steel trusses. Structural tees are created by cutting an S or W beam along the length of the beam's web to create a T shape. A structural tee cut from a wide-flange beam is designated by "WT" followed by its nominal depth and weight per foot. A structural tee cut from an American standard beam is designated by "ST" followed by its nominal depth and weight per foot.

Angles (L) are used as cross-bracing, lintels, ledgers, and connectors and in the construction of steel trusses. Angles are designated by an "L" followed by the length of its longest leg, the length of the other leg, and the thickness of the angle.

Structural Tubing is used as columns and cross-bracing and in the construction of steel trusses. Structural tubing is designated by the length of its longest cross-sectional axis, the length of the other axis, and its wall thickness.

Pipe is used for columns. Pipe is designated by "Pipe" followed by the nominal diameter of the pipe and the type of pipe (e.g., Pipe 4 Std.)

Plate steel is used as base plates and connectors. Plate steel is designated by the thickness.

The weight of a steel member must be determined to estimate the price for the member and to select the correct size of crane. The weight of a steel member is determined by multiplying the length of the steel member by the weight per foot. For wide-flange beams, American standard beams, American standard channels, and structural tees, the weight per foot is obtained from the designation. For angles, structural tubing, pipe, and plate, the weight must be looked up in The *Manual of Steel Construction* published by the American Institute of Steel Construction Inc. (AISC) or another reference that includes excerpts from this manual. The weight of plate steel is about 490 pounds per cubic foot.

Example: Determine the weight of a 25-foot-long W16×36.

$$\text{Weight} = (25 \text{ ft})(36 \text{ lb/ft}) = 900 \text{ lb}$$

Common Steel Connection Types

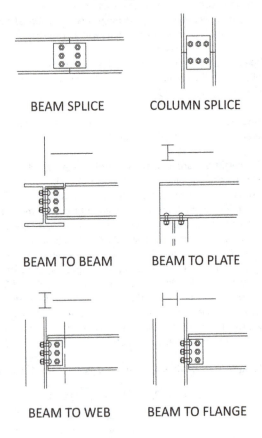

BEAM SPLICE   COLUMN SPLICE

BEAM TO BEAM   BEAM TO PLATE

BEAM TO WEB   BEAM TO FLANGE

Joist and Deck

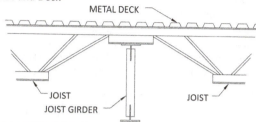

METAL DECK

JOIST
JOIST GIRDER
JOIST

Joist and Deck framing systems are used to support the roof or a concrete floor and consist of a **Metal Deck** supported by **Metal Joists**. Standard metal joists are designed to support a uniform load from the metal deck and are designated as follows:

DEPTH — 24 K 3
JOIST TYPE
SECTION NUMBER

Custom joists may be designed for concentrated and uniform loads. Joists are supported by walls, steel beams, or **Joist Girders**. Joist girders are designated as follows:

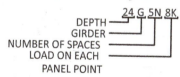

DEPTH — 24 G 5N 8K
GIRDER
NUMBER OF SPACES
LOAD ON EACH
PANEL POINT

The costs and weights of joist and joist girders are obtained from the manufacturer. Visit www.steeljoist.org for more information on steel joists.

Wood Framing

Standard Lumber Sizes

Nominal dimensions are used to specify the width and thickness of lumber. The actual width and thickness of the lumber is smaller than the nominal width and thickness. The nominal and actual sizes are shown in the following table.

Nominal Size	Actual Size
1" x 2"	3/4" x 1-1/2"
1" x 3"	3/4" x 2-1/2"
1" x 4"	3/4" x 3-1/2"
1" x 5"	3/4" x 4-1/2"
1" x 6"	3/4" x 5-1/2"
1" x 7"	3/4" x 6-1/4"
1" x 8"	3/4" x 7-1/4"
1" x 10"	3/4" x 9-1/4"
1" x 12"	3/4" x 11-1/4"
2" x 2"	1-1/2" x 1-1/2"
2" x 3"	1-1/2" x 2-1/2"
2" x 4"	1-1/2" x 3-1/2"
2" x 6"	1-1/2" x 5-1/2"
2" x 8"	1-1/2" x 7-1/4"
2" x 10"	1-1/2" x 9-1/4"
2" x 12"	1-1/2" x 11-1/4"
3" x 6"	2-1/2" x 5-1/2"
4" x 4"	3-1/2" x 3-1/2"
4" x 6"	3-1/2" x 5-1/2"

Framing Styles

With **Balloon Framing** the exterior wall studs run from the foundation to the roof, passing through the floors, as shown in the figure on the left.

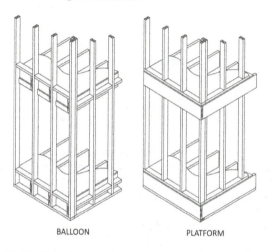

BALLOON

PLATFORM

With **Platform Framing** the exterior wall studs run from the top of one floor to the underside of the floor above it, as shown in the figure on the right.

Floor Framing

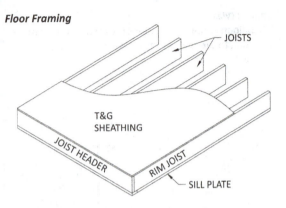

Wood-framed floors consist of **Plywood** or **Oriented Strand Board (OSB)** floor sheathing supported by 2× wood joists or **Engineered Wood I-Joists** (such as TJI). The joists are supported by wood walls, steel beams, **Wood Beams**, or **Sills** that sit on concrete or masonry walls. The beams may sit on steel or **Wood Columns**.

The **Floor Sheathing** is installed with the long direction of the sheet running perpendicular to the joists, with the short side of the sheet nailed to the joists. The floor sheathing usually has a tongue running along one of its long sides and a matching groove running along the other. The tongue and groove are fitted together on the side of the sheathing not supported by the joists and helps to stiffen the floor and reduce floor squeaks. When installing sheathing, waste material may only be used when it has the required tongue and groove.

172

The most accurate way to estimate sheathing is to sketch a layout of the sheathing. Alternatively, for a rectangular floor one can estimate number of rows of sheathing by dividing the width of the floor by the width of the sheathing (4 feet) and rounding up to the next whole number. The number of sheets per row is estimated by dividing the length of the floor by the length of the sheathing (8 feet) and rounding up to the next whole, half, third, or quarter of a sheet. The number of sheets required is the number of rows multiplied by the number of sheets per row. For an irregularly shaped floor, the floor may be divided into a number of rectangles.

Joists are typically run in the direction that minimizes the distance between supporting walls and beams. When more than one joist is needed to get the required length, the joints are placed over the supporting members. One of the advantages of I-joists is that they are available in lengths up to 60 feet, reducing the number of joints. Additional joists may be added under walls that sit on the floor.

Joist Headers, constructed of 2× lumber, I-joists, or other engineered lumber, are placed perpendicular to the joists at the edge of the floor to keep the joists in place and to prevent them from rolling onto their sides. **Cross Bracing** (which is installed in pairs) or **Solid Blocking** is placed anytime the joists pass over a supporting member (except when there is a joist header) and may be placed at specified intervals along the joists to prevent the joists from rolling. **Rim Joists** are joists that are placed parallel to the joist at the edge of the floor.

Openings in floors consist of **Joist Trimmers**, one or more joists that are placed along the side of the opening, **Joist Headers**, that are placed along the edge of the opening perpendicular to the joists, and **Tail Joists**, short joists that span from the joist header to a bearing wall or beam. The joist trimmers and tail joists are usually constructed of the same materials as the joists. The joist headers may be constructed of the same materials as joist (which is often doubled for openings wider than 4 feet) or engineered lumber.

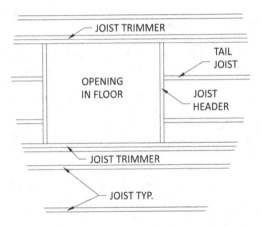

Framing Hardware, such as Simpson joist hangers, may be used to connect joist headers around openings to joists, tail joists to joist headers around openings, and joists to beams.

The length of the joists is determined by the distance between the supporting members and is rounded up to the next standard size. The number of **Joists** is estimated by dividing the length of the building by the spacing of the joists and adding one for a starting joist. If there is more than one row of joists, the number of joists is calculated for each row. Additional joists are added for joists under walls, **Joist Trimmers** and **Joist Headers** around openings in the floor, and for doubled **Rim Joists**. The joist calculations already includes one rim joists at each end. Shorter joists may be ordered for the **Tail Joists**. The quantity of **Joist Headers** at the edges is twice the length of the floor and rounded up the next standard size.

The material needed for **Solid Blocking** is equal to the length of floor multiplied by the number of rows of blocking. When calculating the quantity of solid blocking, the thickness of the joists is ignored. The quantity of **Cross Bracing** per row is equal to two times the length of the floor divided by the joist spacing. The cross bracing for all of the rows is added together to get the total quantity of cross bracing needed.

Framing Hardware is estimated by counting the quantity needed.

The joist may be supported by wood **Sills** sitting on concrete or masonry walls. Sills that come into contact with concrete or masonry must be made of naturally durable wood (e.g., redwood) or must be treated to prevent decay and rot. The sills may be estimated by the individual pieces needed or by the total lineal feet.

The joists may be supported by **Wood Beams**. The length of the wood beams is determined by the distance between the supporting members and is rounded up to the next standard size. Wood beams are estimated by counting the number of pieces of wood needed for the beams.

Wood Columns may be used to support wood beams. The quantity of lumber needed for the columns is determined by counting the columns, noting their size and length.

Example: Determine the materials needed for the floor in the following figure. The joists are 11 7/8 I-joists and the joist headers are 1 3/4×11 7/8 engineered lumber. The joists sit on 2×4 sills on concrete walls. The floor is sheathed with 23/32 OSB.

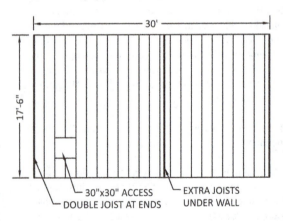

Treated sill will be needed around the perimeter of the building. Order a 16 foot and a 14 foot for each of the 30-foot-long side of the floor and order a 10 foot and an 8 foot for each of the ends.

Order 18-foot-long joists for all of the joists. The two tail joists will be cut from a single joist.

$$\text{Joists} = \frac{(30 \text{ ft})(12 \text{ in/ft})}{(16 \text{ in})} + 1 = 24 \text{ ea}$$

Add one joist for each end and add two joists for under the wall, bringing the total to 28 joists. Order two 33-foot-long pieces of engineered lumber for the joist headers. Three feet will be cut off of each piece for the joist header around the opening.

The sheathing is calculated as follows:

$$\text{Rows} = \frac{17.5 \text{ ft}}{4 \text{ ft}} = 4.375 \text{ round up to } 5$$

$$\text{Sheets per row} = \frac{30 \text{ ft}}{8 \text{ ft}} = 3.75 \text{ round up to } 4$$

$$\text{Sheets} = (4)(5) = 20 \text{ sheets}$$

Stairs

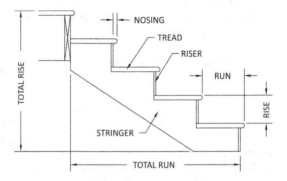

Stairs consist of **Treads** and **Risers** supported by two or more **Stringers**. The treads and risers may be made of wood or particle board. The Stringers are made of 2×12 lumber or engineered lumber. Typically the maximum rise between treads is 7 inches for commercial stairs and 7 ¾ inches for residential stairs. Typically, the minimum length of the tread, measured from the nose of one tread to the nose of the tread above, is 11 inches for commercial stairs and 10 inches for residential stairs.

The number of **Risers** for a flight of stairs may be estimated by dividing the total rise (the vertical distance between floors) by the maximum riser height and rounding up. If you round down you will exceed the maximum riser height. The number of **Treads** is one less than the number of risers for a flight of stairs. The total run (the length of the stairs) for a flight of stairs is equal to the number of treads multiplied by the length of the treads. The length of the **Stringers** can be approximated using the following equation:

$$\text{Length} = \sqrt{\left(\text{Total rise}\right)^2 + \left(\text{Total run}\right)^2}$$

Example: Determine the number of risers and treads and the length of the stringers needed for a flight of stairs between two floors with a total rise of 9 feet. The maximum rise per stair is 7 ¾ inches and the minimum tread length is 10 inches.

$$\text{Risers} = \frac{\left(9 \text{ ft}\right)\left(12 \text{ in/ft}\right)}{\left(7.75 \text{ in/ea}\right)} = 13.9 \text{ ea} \quad \text{Round up to 14 ea}$$

$$\text{Treads} = 14 \text{ ea} - 1 \text{ ea} = 13 \text{ ea}$$

$$\text{Total run} = \frac{\left(13 \text{ ea}\right)\left(10 \text{ in/ea}\right)}{\left(12 \text{ in/ft}\right)} = 10.83 \text{ ft}$$

$$\text{Length} = \sqrt{\left(9 \text{ ft}\right)^2 + \left(10.83 \text{ ft}\right)^2} = 14.08 \text{ ft}$$

Order 16-foot-long stringers.

Wall Framing

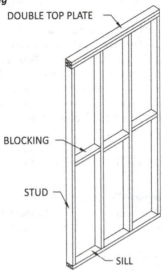

DOUBLE TOP PLATE

BLOCKING

STUD

SILL

Wood framed walls consist of **Studs** nailed between a **Sill** (bottom) plate and a **Top Plate**. Sill plates that are in contact with concrete or masonry must be made of naturally durable wood (e.g., redwood) or must be treated to prevent decay and rot. In bearing walls the top plate typically consists of two plates with their joints staggered. Blocking is often included in walls to prevent the spread of fire (**Fire Blocking**), as backing for handrails or other items (**Backing**), or to strengthen a bearing wall.

The quantity of **Plate** needed equals the number of plates times the length of the wall and may be estimated by the individual pieces needed or by the total lineal feet. Treated and redwood plates must be kept separate from the ordinary plate.

The number of **Studs** may be estimated by dividing the length of the wall by the stud spacing and adding studs for corners, wall intersections, and door and window openings.

Two common methods of framing **Corners** are shown in the following figure. Typically three additional studs are needed for each corner: one stud to end the wall at a point other than the standard stud spacing, one for backing on the inside of the corner, and one to start the new wall.

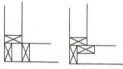

Two common methods of framing **Intersections** are shown in the following figure. Typically three additional studs are needed for each intersection: two for backing on the inside of the corners and one to start the new wall.

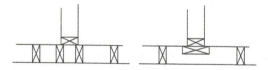

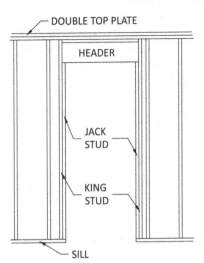

Text labels in image: DOUBLE TOP PLATE, HEADER, JACK STUD, KING STUD, SILL

Door Openings consist of a **Header** supported by **Jack Studs**. A **King Stud** is placed alongside of the jack stud and header to tie the opening together. In bearing walls, the header may be made of 2× wood with OBS spacers, solid wood, or engineered lumber. For non-bearing walls, the header may be a single stud. Wider openings may require two or three jack studs on each side of the opening. The length of the header is equal to the width of the opening plus 1 ½ inches for each jack stud. The opening must be wide enough for the door, the door frame, and space for shims.

The actual number of additional studs needed for a door opening varies based upon the location and width of the door. Typically two studs are added for each door opening.

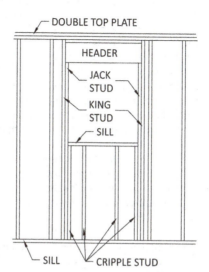

DOUBLE TOP PLATE

HEADER

JACK STUD

KING STUD

SILL

SILL — CRIPPLE STUD

Window Openings are framed the same as door openings, except the bottom of the opening is filled with **Cripple Studs** supporting the window **Sill**.

The actual number of additional studs required for window opening varies based upon the window location and the window height. For short windows, only one cripple stud can be cut from a regular stud. Short windows typically require seven additional studs (two king studs, two jack studs, two cripple studs next to the jack studs, and a sill). The number of studs may be reduced if any of these studs falls on the standard stud spacing. For taller windows, where two or more cripple studs may be cut from a single stud, the number of additional studs may be reduced.

Exterior walls and interior shear walls are often sheathed with **Plywood** or **OSB** sheathing. For walls with a consistent height, the number of rows of sheathing may be estimated by dividing the wall height by the height of the sheathing and rounding up to the next whole, half, third, quarter, etc. of a sheet. The number of sheets per row is estimated by dividing the length of wall by the width of the sheathing. The number of sheets required is the number of rows multiplied by the number of sheets per row. Waste must be added for materials that are left over from one wall and are too small to be used on another wall. For raked walls—walls with a sloped top— the best way estimate sheathing is to sketch a layout of the sheathing.

Example: Determine the materials needed to frame the exterior walls of a 16-foot by 22-foot storage shed. The walls are 8-feet high and are framed with 2×4 lumber at 16 inches on center. There are two 24-inch by 24-inch windows and one 36-inch-wide by 80-inch-high door. The exterior of the walls are sheathed with 7/16-inch OSB. The floor of the shed is concrete.

The sills will need to be treated. Order a 14 foot and an 8 foot for each of the 22-foot-long side and order a 16 foot for each of the 16-foot-long sides. Order twice as much non-treated plate for the top plates. Order the following quantities:

- 2 each 2×4 – 8' Treated
- 2 each 2×4 – 14' Treated
- 2 each 2×4 – 16' Treated
- 4 each 2×4 – 8'
- 4 each 2×4 – 14'
- 4 each 2×4 – 16'

The number of studs needed is calculated as follows:

$$\text{Length of wall} = 16 \text{ ft} + 22 \text{ ft} + 16 \text{ ft} + 22 \text{ ft} = 76 \text{ ft}$$

$$\text{Studs} = \frac{(76 \text{ ft})(12 \text{ in/ft})}{(16 \text{ in})} + (4 \text{ corners})(3 \text{ ea/corner})$$
$$+ (2 \text{ windows})(7 \text{ ea/window})$$
$$+ (1 \text{ door})(2 \text{ ea/door})$$

$$\text{Studs} = 85 \text{ ea}$$

One row of sheathing is needed.

$$\text{Sheet per row} = \frac{76 \text{ ft}}{4 \text{ ft}} = 19 \text{ sheets/row}$$
$$\text{Sheets} = (19 \text{ sheets/row})(1 \text{ row}) = 19 \text{ sheets}$$

Stick-Framed Roofs

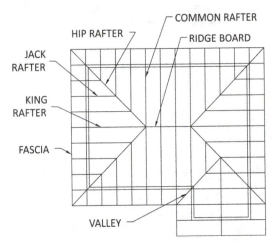

Stick-framed roofs are framed with 2× lumber or I-joist rafters. **Common Rafters** run from the ridge to the eaves. **Hip Rafters** are located at the junction of two inclined roof planes and create a peak. **Valley Rafters** are located at the junction of two inclined roof planes and form a valley. On roofs where all of the planes have the same slope, the slope of the hip and valley rafters is determined by replacing the 12 in the slope with 17 (e.g. on roof with a 4:12 slope the hip rafter will have a 4:17 slope). **King Rafters** run from the intersection of two hip rafters to the eave of the roof. **Jack Rafters** are shorter rafters that connect to a hip or valley rafter.

A Ridge Board is placed at the ridge of the roof to support the rafters and is often constructed of 1× lumber, 2 inches wider than the rafters. When a stick-framed roof is loaded, the rafters create an outward force on the wall near the eaves that support them. **Ceiling Joists** or **Collar Ties** are required to prevent the walls from spreading. Ceiling joists are used to support a flat ceiling. Collar ties connect two rafters on opposite sides of the ridge together. Collar ties are usually placed about 1/3 of the way from the ridge to the bearing wall at the eaves and are spaced no more than 48 inches apart. They are often constructed of 1× lumber and have a typical length equal to 1/3 of the span.

Rafters are estimated in the same manner as joists. Because the rafters are sloped, the length shown on the plans must be converted to the actual length by one of the following equations:

$$\text{Length} = \sqrt{\left(\text{Rise}\right)^2 + \left(\text{Run}\right)^2}$$

$$\text{Length} = \left(\text{Length}_{\text{Plan view}}\right)\sqrt{1 + \left(\frac{\text{Slope}}{12}\right)^2}$$

Ceiling joists are estimated in the same manner as joists. Collar ties are estimated by dividing the length of the ridge by the spacing of the collar ties and adding one as a starting collar tie. The collar tie spacing must be a multiple of the rafter spacing.

Example: Determine the materials needed to frame a 20-foot-wide by 24-foot-long gable-end roof with a slope of 4:12. The rafters are constructed from 2×6 lumber at 16 inches on center, the ridge board is constructed of 1×8 lumber, and the collar ties are constructed of 1×4 lumber.

The number of rafters on one side of the roof is calculated as follows:

$$\text{Rafters} = \frac{(24 \text{ ft})(12 \text{ in/ft})}{(16 \text{ in})} + 1 = 19 \text{ ea}$$

Thirty-eight (2×19) rafters are required. Their plan view length is 10 feet (20 ft/2). The rafter length is calculated as follows:

$$\text{Length} = (10 \text{ ft})\sqrt{1 + \left(\frac{4}{12}\right)^2} = 10.5 \text{ ft} \quad \text{order 12 ft}$$

Order thirty-eight 12-foot-long 2×6s. The ridge is 24 feet long. Order two 12-foot-long 1×8s for the ridge. The collar ties need to be 6.7 feet (20 ft/3) long. The number of collar ties is calculated as follows:

$$\text{Collar ties} = \frac{(24 \text{ ft})(12 \text{ in/ft})}{(48 \text{ in})} + 1 = 7 \text{ ea}$$

Order seven 8-foot-long 1×4s for the collar ties.

Soffit and Fascia

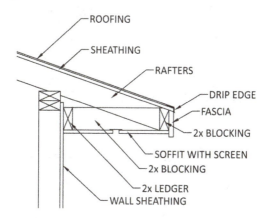

Fascia runs around the edge of the roof covering the end of the rafters. The fascia may be wood trim or aluminum or vinyl over rough lumber. The underside of the overhang of the roof is often covered with soffit. The **Soffit** may be a finish-grade plywood or aluminum or vinyl soffit. The soffit is vented to allow air to circulate through attic.

Example: Determine the lineal feet of soffit and fascia needed for the roof in the previous example.

The length of the soffit and fascia on the gable ends is the same as the length of the rafters. The length of the soffit and fascia is equal to the four times the length of the rafters plus two times the length of the building.

$$\text{Length} = 4(10.5 \text{ ft}) + 2(24 \text{ ft}) = 90 \text{ ft}$$

189

Trusses

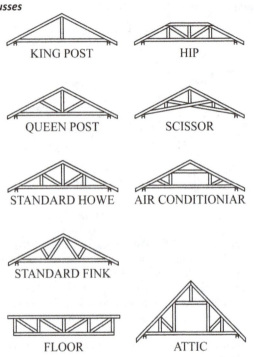

KING POST

HIP

QUEEN POST

SCISSOR

STANDARD HOWE

AIR CONDITIONIAR

STANDARD FINK

FLOOR

ATTIC

Trusses need to be designed for the snow and wind loads for a specific location. Bids for trusses should be obtained from a truss manufacturer.

Insulation

Insulation is used to resist heat flow from warm to cold. In the winter it resists heat loss from the building and in the summer it resists heat gain into the building. Insulation's R-value is a measurement of how well it resists heat transfer. The higher the R-value the better it resists heat transfer. Building codes set the minimum R-value for the walls, roofs, and floor of buildings.

Batt Insulation consists of blankets and batts of fiberglass, mineral wool, or other fibers and may be faced with Kraft paper. Batt insulation is placed between studs or joists and is one inch narrower than the stud or joist spacing, which allows it to fit snugly between the framing members. When batt insulation is faced with Kraft paper, the paper is stapled to the framing members. Unfaced batt insulation may be covered with a sheet of plastic on the inside face of the insulation, which acts as a moisture barrier. The quantity of batt is calculated using the following formula.

$$\text{Quantity} = \text{Area}\left(\frac{\text{Spacing} - 1\ \text{in}}{\text{Spacing}}\right)$$

Example: Determine the quantity of insulation needed for a 100-foot-long by 8-foot-high wall. The studs are spaced 16 inches on center.

$$\text{Quantity} = \left(100\ \text{ft}\right)\left(8\ \text{ft}\right)\left(\frac{16\ \text{in} - 1\ \text{in}}{16\ \text{in}}\right) = 750\ \text{ft}^2$$

Loose Insulation consists of cellulose (wood fibers or shredded newspapers chemically treated to be fire and insect resistant), fiberglass, miner wool, or other fiber that is blown in place using special equipment or poured in place. Loose insulation works well in irregularly shaped areas or for filling existing framed walls. It works poorly on sloped ceilings because it can settle, leaving parts of the ceiling exposed. The R-value of loose insulation is dependent on how thick of a layer of insulation is placed. The quantity of loose insulation is calculated based upon the number of bags needed to cover 1,000 square feet and is often printed on the insulation's packaging.

Example: Determine the number of bags of loose insulation needed for a 1,600-square-foot ceiling. Twenty-nine bags are needed to cover 1,000 square feet.

$$\text{Quantity} = \left(1,600 \text{ ft}^2\right)\left(\frac{29 \text{ Bags}}{1,000 \text{ ft}^2}\right) = 47 \text{ Bags}$$

Rigid Foam Insulation consists of sheets of polystyrene, polyurethane or other foam and is applied to the surface of walls and ceilings. Rigid foam insulation is estimated in the same way as sheathing is estimated.

Sprayed in Foam insulation consists of polyurethane or other foam that is sprayed between the framing members and has the added advantage of sealing possible sources of air infiltration.

Roofing

Roof Flashings

Flashings are required around the edges of the roof, roof penetrations, and where the roof intersects with walls. The most common material for roof flashings is metal. Plastic flashings are often used around plumbing pipes.

Shingle/Shake Roofs

Shingle roofs consist of **Asphalt Shingles** or **Western Red Cedar Shingles/Shakes** installed over **Asphalt-Impregnated Felt**. The shingles/shakes are nailed to the underlying roof sheathing. **Starter Rows** of shingles are required at the bottom edges of the roof and often special **Cap Shingles** are used on the ridges and hips. Shingles are measured by the square (sq), which is the quantity of shingles need to cover 100 square feet of roof and is based upon a standard exposure (the distance between rows of shingles). The following equation takes the slope of the roof into account:

$$\text{Quantity} = \frac{\left(\text{Area}_{\text{Plan view}}\right)}{\left(100 \text{ ft}^2/\text{sq}\right)} \sqrt{1 + \left(\frac{\text{Slope}}{12}\right)^2}$$

Example: Determine the number of square of shingles need for a 40-foot-wide by 50-foot-long roof. The slope on the roof is 5:12.

$$\text{Quantity} = \frac{\left(40 \text{ ft}\right)\left(50 \text{ ft}\right)}{\left(100 \text{ ft}^2/\text{sq}\right)} \sqrt{1 + \left(\frac{5}{12}\right)^2} = 21.7 \text{ sq}$$

Tile Roofs

Tile roofs consist of **Slate** (natural stone) or **Clay Tiles** installed over **Asphalt-Impregnated Felt**. The tiles are nailed to the underlying roof sheeting. The tiles may be flat or curved. Curved tiles require a **Bird Stop** tile at the bottom edges of the roof. Special cap tiles and mortar are used at the gable ends, ridges, and hips. Tile roofs are used on roofs with a slope of at least 4:12.

Built-up Roofs

Built-up roofs consist of layers (plies) of roofing felts adhered together with hot asphalt. The durability of the roof is determined by number of plies. Gravel is often embedded in the last layer of asphalt to protect the roof and provide a reflective surface that resists heat gain. Build-up roof are used on nearly flat roofs.

Single-Ply Membrane Roofs

Single-ply membrane roofs consist of large sheets of plastic membrane welded together. The sheets are made from **EPDM** (rubber), polyvinyl chloride (**PVC**), or another combination of plastics. The membrane is fastened at the edges with **Batten Strips**. The center of the sheet is fastened in place or held in place with rock ballast. Single-ply membrane roofs are used on nearly flat roofs.

Metal Roofs

Metal roofs consist of sheets of metal installed over **Asphalt-Impregnated Felt** and are fastened to the roofing deck. Common materials for metal roofs include steel with a painted or baked enamel finish, copper, or aluminum.

Aluminum and Vinyl Siding

Aluminum and vinyl siding consists of a preformed siding panel made from aluminum or vinyl, which is often formed to look like wood, and is installed over a building paper, such as **Asphalt-Impregnated Felt** or **Tyvek**. The siding has interlocking lips on the top and bottom. The bottom lip locks into the top lip of the previous row or the **Starter Strip** in the case of the bottom row. The top lip is nailed to the wall sheathing or framing. **J-Molding** is required around doors, windows, at the top of the walls, and on inside corners. **Outside Corner** pieces are required on the outside corners. The quantity of siding is measured in squares (the amount of siding needed to cover 100 square feet). Starter strips, J-moldings, and outside corners are bid by the piece.

Matching **Soffit** and **Fascia** is often available. The top of the fascia tucks under the drip edge of the roof and the bottom edge covers the soffit. The soffit requires a **J-Molding** at the wall side. Narrow pieces of soffit are held in place by the J-molding and fascia. Wider pieces of soffit must be nailed to the underlying framing members.

Rain Gutters

Rain Gutters are placed at the bottom edge of the roof to catch and direct perception from the roof to the ground. Rain gutters may be made from aluminum, finished steel, copper, or polyvinyl chloride (PVC). **Downspouts** are pipes that drain the water from the gutters to the ground.

Doors

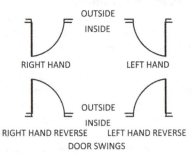

OUTSIDE
INSIDE

RIGHT HAND   LEFT HAND

OUTSIDE
INSIDE

RIGHT HAND REVERSE   LEFT HAND REVERSE
DOOR SWINGS

Doors consist of the **Door Slab**, **Frame**, and **Hardware** and are bid by counting the required components.

The types of door slabs include **Hollow-Core Wood** (two sheets of wood or hardboard with a wood edge and hollow center), **Solid-Core Wood**, **Steel** (two sheets of steel with a wood edge filled with insulation), **Fiberglass** (two sheets of fiberglass with a wood edge filled with insulation), and **Hollow Metal** (a steel door with a hollow center).

Frame types include **Solid Wood** and **Hollow Metal** (steel). The top of the frame is known as the **Head**; and the sides are known as the **Jamb**.

Door hardware include **Butts** (hinges), **Weather Stripping**, **Threshold** (between the door and the floor), **Lockset** (door handle), **Deadbolt**, **Stop** (limits the swing of the door), **Panic Hardware**, and **Closer**.

Storefronts

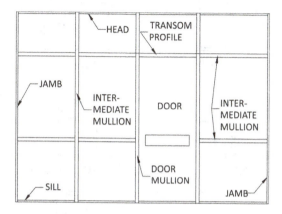

Storefronts consist of **Glass** set in frames of **Aluminum Tubing** forming large glazed areas which may include doors. Storefronts are often used in the front of stores and commercial buildings, allowing retailers to display their wares. Storefronts are usually bid by obtaining a price from a subcontractor.

Windows

FIXED

SLIDING

SINGLE HUNG

DOUBLE HUNG

CASEMENT

AWNING

Windows consist of a single assembly and are used primarily in residential and light-commercial construction. The frames of windows may be made from wood, vinyl, vinyl covered wood, aluminum, or steel. The top of the window is known as the **Head**, the sides are known as the **Jamb**, and the bottom is known as the **Sill**. The window may include single, double, or triple panes of tinted or **Low-E Glass** and the space between the panes may be filled with **Argon** or **Krypton** gas to decrease heat transfer through the window. For hinged windows, the intersection of the dashed lines shows which side is hinged.

Light-Gage Metal Framing

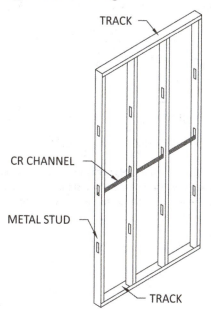

TRACK

CR CHANNEL

METAL STUD

TRACK

Light-gage metal framing is used to frame non-bearing walls and consists of a **Metal Stud** fastened with screws to a top and bottom C-shaped **Track**. Track is also used for the headers. Cold rolled steel (CR) channel may be run through the preformed holes in the metal studs to stiffen the walls. These same materials may be used to frame ceiling drops and other features. Light-gage metal framing is bid the same as wood framing, except the studs run from the bottom of the bottom plate to the top of the top plate.

Example: What materials are needed to frame 20 feet of metal stud partition? The studs are 16 inches on center.

$$\text{Studs} = \frac{(20 \text{ ft})(12 \text{ in/ft})}{(16 \text{ in})} + 1 = 16 \text{ ea}$$

Forty feet of track and 16 studs.

HAT CHANNEL   RC1 CHANNEL

Metal furring, such as a **Hat Channel**, may be fastened to concrete and masonry to provide support for the drywall. **RC1 Channel** is used between the gypsum board and the framing to increase the sound resistance of the wall or ceiling.

Drywall

Drywall consists of sheets of **Gypsum Board** (gypsum plaster dried between the sheets of thick paper) fastened by nails, screws, or glue to the wood or metal framing of walls and ceilings. **Type-X** gypsum board is used in fire-rated construction. **Water-Resistant Gypsum Board** must be used in wet or damp locations. Gypsum board is commonly available in 4-foot-wide sheets ranging in length from 8 to 16 feet and in thicknesses of 1/4, 1/2, 5/8, and 1 inch. On ceilings the long direction of the sheet is run perpendicular to the joists. On walls the sheet may be run horizontal or vertically. Metal trim is used on exterior corners (**Corner Bead**) and edges (**J-Molding**). The seams are concealed and the trim incorporated into the wall by taping them with **Joint Compound** (mud). **Paper Joint Tape** is taped in the joints as reinforcing. Multiple layers of joint compound are applied and sanded to provide a smooth wall. Drywall may bid the same way as sheathing. More commonly it is bid by a subcontractor.

Example: Determine the drywall needed for an 8-foot-high by 24-foot-long wall. The drywall is run horizontally and is available in 8, 10, 12, 14, and 16 foot lengths. To finish 100 square feet of drywall, 0.4 boxes of joint compound (mud) and 0.13 rolls of joint tape are needed.

Order four 12-foot-long sheets.

$$\text{Area} = (8 \text{ ft})(24 \text{ ft}) = 192 \text{ sf}^2$$

$$\text{Mud} = (192 \text{ sf}^2)(0.4 \text{ boxes}/100 \text{ sf}^2) = 0.8 \text{ boxes}$$

$$\text{Tape} = (192 \text{ sf}^2)(0.13 \text{ rolls}/100 \text{ sf}^2) = 0.25 \text{ rolls}$$

Acoustical Ceilings

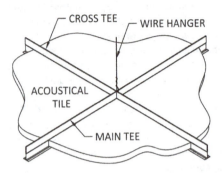

CROSS TEE — WIRE HANGER

ACOUSTICAL TILE

MAIN TEE

Acoustical ceilings consist of **Acoustical Tiles** set in a metal or PVC grid. The grid is suspended on wires and consists of **Main Tees** running the full width or length of the room at four feet on center and 2- and 4-foot-long **Cross Tees** which break the grid down into 2-foot by 4-foot or 2-foot by 2-foot squares. A **Wall Angle** is used to finish the edges of the room. A **Seismic Brace** is needed for every 100 square feet of ceiling, and consists of four wires and a solid rod that keep the ceiling from swaying and bouncing up and down. Some of the acoustical tiles are replaced with **Light Fixtures**, **HVAC Grills**, and **Decorative Panels**.

Acoustical ceilings are often bid by the square foot. The most accurate way to estimate acoustical ceilings is to prepare a layout, deducting tiles that have been replaced with light fixtures, HVAC grills, and decorative panels.

Ceramic Tile

Ceramic tile consists of pieces of ceramic tile adhered to the walls, floors, or ceilings. With **Thickset** or **Mud Set** tile, a layer of mortar is applied to the substrate and the tile is adhered to the mortar. Applying the mortar is a time consuming process. With **Thinset** tile, the tile is adhered to a backer board, which is nailed or screwed to the substrate. **Water-Resistant Gypsum Board** may be used as a backer board on walls and ceilings where exposure to moisture is limited, such as backsplashes in residential kitchens and bathrooms and wainscots in bathrooms. **Cementitious Backer Boards** should be used when the tile is used in high-moisture areas, such as in shower stalls, and on countertops and floors. The space between the tiles is filled with **Grout**. Specialty tiles, such as **Corner Bullnose** may be used at the edges to provide a finished surface. **Decorative Tiles** may be laid into the pattern.

A common way to estimate the number of tiles needed is to divide the area of the floor by the area of the tile and add 10% for waste. The most accurate way to estimate tile is to prepare a layout. Specialty and decorative tiles must be kept separate in the estimate.

Example: Determine the number of 6-inch by 6-inch tiles needed to finish the floor of a 5-foot by 6-foot restroom. Add 10% for waste.

$$\text{Quantity} = \frac{(4 \text{ ft})(12 \text{ in/ft})(6 \text{ ft})(12 \text{ in/ft})}{(6 \text{ in})(6 \text{ in})}\left(1 + \frac{10}{100}\right)$$

$$\text{Quantity} = 106 \text{ tiles}$$

Flooring

Wood Flooring consists of strips (up to 3.25 inches wide) or planks (3.25 to 8 inches wide) of solid wood or engineered wood (a finished wood surface layer bonded to a core of thin plies of wood or fiberboard to provide stability). The flooring strips/planks are connected to each other by a tongue and grove, snap lock, or spline joint. The wood floor may be glued or nailed to the substrate or it may float over the substrate allowing the floor to expand and contract with changes in temperature and humidity. Unfinished wood floors are sanded and may be stained before applying a varnish, urethane, or wax finish. Some wood floors come prefinished. Wood floors are estimated by dividing the area of the floor by the coverage in a box or bundle of flooring and adding waste.

Example: How many bundles of wood flooring are needed for a 10-foot by 10-foot room? There is 20.5 square feet per bundle. Add 10% waste.

$$\text{Quantity} = \frac{(10 \text{ ft})(10 \text{ ft})}{(20.5 \text{ ft}^2/\text{box})}\left(1 + \frac{10}{100}\right) = 5.4 \text{ boxes}$$

Laminate Flooring consists of a top or wear layer containing aluminum oxide, a pattern layer giving the floor the look of real wood or tile, a fiber board layer that may be impregnated with plastic resin (melamine), and a backing layer to protect against moisture. The planks are connected by a tongue and groove joint that is glued or a snap lock joint. Laminate floors are installed over a pad and float over the floor. Laminate floors are estimated the same way as wood floors.

Carpet consists of nylon, olefin (polypropylene), nylon and olefin blends, acrylic, polyester, recycled PET (polyethylene terephthalate), or wool fibers on a backing material. The primary layer of the backing material is woven through the fibers. The secondary backing layer protects the carpet from moisture, bacteria, and molds. **Residential Carpet** comes in rolls 12 feet wide and is installed over pad. The carpet is secured to the substrate by tackless strips placed around the perimeter of the room. Seams are glued together with seaming tape and a hot seaming iron.

Commercial Carpet is available in 12 foot wide rolls and 2-foot by 2-foot squares. Commercial carpet is glued to the substrate or to a pad that has been glued to the substrate. When roll carpet is installed in a room, the pieces must be installed in the same direction and the pattern must be matched. Transition strips may be used where the carpet meets other flooring materials. To properly estimate the quantity of carpet needed, a layout of the carpet in the rooms (a seaming plan) must be made. The direction in which the carpet is installed can change the required quantity of carpet greatly.

Example: How many square yards of carpet is needed for a 20-foot by 18-foot room? The carpet comes in rolls 12 feet wide.

Two 12-foot-wide by 18-foot-long pieces are needed.

$$\text{Quantity} = (2 \text{ ea}) \frac{(12 \text{ ft})(18 \text{ ft})}{(9 \text{ ft}^2/\text{yd}^2)} = 48 \text{ yd}^2$$

Sheet Vinyl consists of a clear protective layer, a pattern layer giving the floor the look of wood or tile, and a backing layer. Sheet vinyl comes in rolls 6 or 12 feet wide and is glued to the substrate or taped at the seams and allowed to float over the substrate. When more than one piece of sheet vinyl is installed in a room, the sheets must be installed in the same direction and the pattern must be matched. Transition strips and moldings are required in doorways, at tubs, and where the sheet vinyl meets another flooring material. A layout of the sheet vinyl (a seaming plan) must be made to properly estimate the quantity of vinyl. Sheet vinyl is estimated in the same way as carpet.

Vinyl Composition Tile (VCT) is composed of colored chips of vinyl formed into tiles 12 inches by 12 inches, which are glued to the substrate. A variety of thicknesses are available, with 1/8 of an inch being the most common. VCT is a good choice in high traffic areas and is used mostly on commercial projects. A common way to estimate the number of tiles needed is to divide the area of the floor by one square foot and add a percentage for waste. The most accurate way to estimate VCT is to prepare a layout, particularly when there is a pattern.

Example: Determine the number of tiles needed to finish a 7-foot by 60-foot hallway. Add 5% for waste.

$$\text{Quantity} = \frac{(7 \text{ ft})(60 \text{ ft})}{(1 \text{ ft}^2/\text{tile})}\left(1 + \frac{5}{100}\right) = 441 \text{ tiles}$$

Paint

Paint consists of a prime coat and one or more coats of finish paint. The prime coat is used to seal the underlying surface, helping the paint bond, and reduce bleed through from the underling materials. The use of primer is necessary when painting bare drywall, wood, metal, or concrete, or over another finish that might bleed through, such as a dark coat of paint. Latex (water based) and alkyd (oil based) paints are used as a finish coat. Paint is available with a flat, satin (eggshell), semi-gloss, and high-gloss finish. Primer and paint may be applied by brush, roller, or sprayer.

Painting of walls, ceilings, and floors is measured by the square foot. Trim that is painted separate from the surround surface is measured by the linear foot. Doors are measured by the door.

Example: Determine the area of the walls and ceilings to be painted in a 12-foot-long by 11-foot-wide by 8-foot-high room. If one gallon of paint covers 350 square feet, how many gallons are required to cover the walls and ceiling with two coats?

$$\text{Wall}_{\text{Length}} = 11 \text{ ft} + 12 \text{ ft} + 11 \text{ ft} + 12 \text{ ft} = 46 \text{ ft}$$

$$\text{Wall}_{\text{Area}} = (46 \text{ ft})(8 \text{ ft}) = 368 \text{ ft}^2$$

$$\text{Ceiling}_{\text{Area}} = (11 \text{ ft})(12 \text{ ft}) = 132 \text{ ft}^2$$

$$\text{Total area} = 368 \text{ ft}^2 + 132 \text{ ft}^2 = 500 \text{ ft}^2$$

$$\text{Gallons} = \frac{(2 \text{ coats})(500 \text{ ft}^2/\text{coat})}{350 \text{ ft}^2 / \text{gallon}} = 3 \text{ gallons}$$

Fire Sprinklers

Automatic fire sprinkler systems are used to extinguish fires before they have a chance to spread. Most fire sprinkler systems use sprinkler heads with a fusible link that melts or a liquid-filled glass vial that bursts when exposed to the heat from a fire, opening the sprinkler head. Fire sprinkler systems are bid by subcontractors. There are four common types of fire sprinkler systems.

In a **Wet Pipe** system, the fire sprinkler pipes are filled with water, which is released when a sprinkler head opens. This is the most common type of system.

In a **Dry Pipe** system, the fire sprinkler pipes are filled with pressurized nitrogen or air, which is released when a sprinkler head opens. The loss in pressure opens a dry-pipe valve allowing water to flow into the sprinkler pipes. A dry pipe system is used in areas where the sprinkler pipe is exposed to freezing temperatures.

In a **Pre-Action** system, water is prevented from entering the piping system by a pre-action valve, which is linked to a fire or smoke detection system. For a pre-action system to operate the pre-action valve must be opened by the fire/smoke detection system and the sprinkler head must be opened by heat.

A **Deluge** system is a pre-action system with sprinkler heads that are open all of the time. When the fire/smoke detection system opens the pre-action valve, all of the sprinkler heads go off. A deluge system is used in high hazard areas.

Plumbing

Plumbing systems include the piping to supply water to the fixtures, the plumbing fixtures (e.g., sinks, water closets, lavatories, etc.), and the drain, waste, and vent (DWV) piping to remove waste water from the building. Plumbing also includes **Medical Gas** piping (vacuum and oxygen) and **Compressed Air**. Plumbing is bid by a specialty contractor.

HVAC

Heating, Ventilation and Air Conditioning (HVAC) systems are responsible for maintaining the air quality in a building. The HVAC system includes equipment to heat, cool, humidify, dehumidify, and filter the air inside a building and fans, supply air ducts, and return air ducts to distribute air throughout the building. It also includes gas (natural gas or fuel oil) piping used to supply energy to heat the building and refrigeration systems such as those used in grocery stores. HVAC is bid by a specialty contractor.

Electrical

The electrical system supplies power throughout the building and includes transformers, switch gear, load panels with circuit breakers, wiring, receptacles, light fixtures, and motors. Electrical is bid by a specialty contractor.

Excavation
Bank, Loose, and Compacted Soils

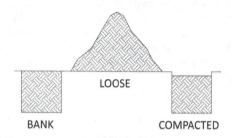

When soil is excavated, it is in its natural or **Bank** condition and is measured in bank cubic yards (bcy). When soil is transported it is in a **Loose** condition and is measured in loose cubic yards (lcy). After fill has been placed and compacted, it is in a **Compacted** state and is measured in compacted cubic yards (ccy).

Swell

When soil is excavated its volume increases as air voids are introduced into the soil. This increase in volume is known as swell and is determined by the following equation, where the density is the density of the dry soil.

$$\text{Swell \%} = \left(\frac{\text{Density}_{\text{Bank}}}{\text{Density}_{\text{Loose}}} - 1 \right) 100$$

Example: A soil in its bank condition has a dry density of 100 pounds per cubic foot. After excavation it has a dry density of 80 pounds per cubic foot. Determine the swell percentage for the soil.

$$\text{Swell \%} = \left(\frac{100 \text{ lbs/ft}^3}{80 \text{ lbs/ft}^3} - 1 \right) 100 = 25\%$$

The swell % is used to convert between bank and loose volumes using the following equation.

$$\text{Volume}_{\text{Loose}} = \text{Volume}_{\text{Bank}} \left(1 + \frac{\text{Swell \%}}{100} \right)$$

Example: One thousand bank cubic yards of soil are to be excavated and hauled off site using 10-cubic-yard trucks. If the swell is 25% how many truck loads are needed?

$$\text{Volume}_{\text{Loose}} = 1,000 \text{ bcy} \left(1 + \frac{25}{100} \right) = 1,250 \text{ lcy}$$

$$\text{Trucks} = \frac{1,250 \text{ lcy}}{10 \text{ lcy/load}} = 125 \text{ loads}$$

Shrink

When soil is compacted, its volume, compared to its original bank condition, decreases (except for blasted rock, which increases). This decrease in volume is known as shrinkage and is determined by the following equation.

$$\text{Shrinkage \%} = \left(1 - \frac{\text{Density}_{\text{Bank}}}{\text{Density}_{\text{Compacted}}} \right) 100$$

Example: A soil in its bank condition has a dry density of 100 pounds per cubic foot. It is placed and compacted to an average density of 105 pounds per cubic foot. Determine the shrinkage percentage for the soil.

$$\text{Shrinkage \%} = \left(1 - \frac{100 \text{ lbs/ft}^3}{105 \text{ lbs/ft}^3}\right)100 = 4.8\%$$

The shrinkage percentage is used to convert between bank and compacted volumes using the following equation.

$$\text{Volume}_{Compacted} = \text{Volume}_{Bank}\left(1 - \frac{\text{Shrinkage \%}}{100}\right)$$

Example: The soil in the above example is to be used as compacted fill on a construction project. How much fill will there be if the soil shrinks 4.8 percent?

$$\text{Volume}_{Compacted} = 1,000 \text{ bcy}\left(1 - \frac{4.8}{100}\right) = 952 \text{ ccy}$$

Typical Swell and Shrink Percentages		
Material	Swell %	Shrink %
Sand and Gravel	10 to 18	0 to 15
Loam	15 to 25	0 to 10
Dense Clay	20 to 30	0 to 10
Solid Rock	40 to 70	-30

Excavation Quantities

The volume of excavation can be calculated by dividing the shape to be excavated into geometric shapes, determining the volume for each shape, and summing their volumes.

Example: Determine the volume of soil that needs to be excavated for a bridge footing. The bottom of the excavation is 10 feet deep by 20 feet wide by 30 feet long. The sides are sloped at a 1:1.

The excavation can be divided into the following shapes:

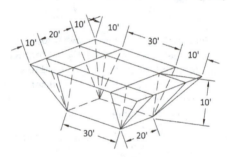

$$\text{Volume}_{\text{Center}} = (20 \text{ ft})(30 \text{ ft})(10 \text{ ft}) = 6,000 \text{ ft}^3$$

$$\text{Volume}_{\text{Side 1}} = (2 \text{ ea})(20 \text{ ft})(10 \text{ ft})(10 \text{ ft})/2 = 2,000 \text{ ft}^3$$

$$\text{Volume}_{\text{Side 2}} = (2 \text{ ea})(30 \text{ ft})(10 \text{ ft})(10 \text{ ft})/2 = 3,000 \text{ ft}^3$$

$$\text{Volume}_{\text{Corner}} = (4 \text{ ea})(10 \text{ ft})(10 \text{ ft})(10 \text{ ft})/3 = 1,333 \text{ ft}^3$$

$$\text{Volume}_{\text{Total}} = 6,000 \text{ ft}^3 + 2,000 \text{ ft}^3 + 3,000 \text{ ft}^3 + 1,333 \text{ ft}^3$$

$$\text{Volume}_{\text{Total}} = (12,333 \text{ ft}^3)\left(\frac{1 \text{ yd}^3}{27 \text{ ft}^3}\right) = 457 \text{ yd}^3$$

Average-End Method

The **Average-End Method** is often used to calculate the excavation volumes for road cuts and fills, but may be used in other types of excavations. The excavation volume is calculated by taking cross sections through the excavation area and determining the area of the cuts and fills for each of the cross sections. The cut or fill between two cross sections is determined by the following formula using the distance between the two cross sections.

$$\text{Volume} = \left(\frac{\text{Area}_1 + \text{Area}_2}{2} \right) \text{Distance}$$

Example: Four sections have been drawn through a road cut. The cuts for Section A and D is 0; the cut for Section B is 75 square feet; and the cut for Section C is 110 square feet. The distance between Sections A and B and Sections B and C is 50 feet. The distance between Section C and D is 20 feet. Determine the volume of the excavation.

$$\text{Volume}_{A\text{-}B} = \left(\frac{0 \text{ ft}^2 + 75 \text{ ft}^2}{2} \right) (50 \text{ ft}) = 1{,}875 \text{ ft}^3$$

$$\text{Volume}_{B\text{-}C} = \left(\frac{75 \text{ ft}^2 + 110 \text{ ft}^2}{2} \right) (50 \text{ ft}) = 4{,}625 \text{ ft}^3$$

$$\text{Volume}_{C\text{-}D} = \left(\frac{110 \text{ ft}^2 + 0 \text{ ft}^2}{2} \right) (20 \text{ ft}) = 1{,}100 \text{ ft}^3$$

$$\text{Volume}_{\text{Total}} = 1{,}875 \text{ ft}^3 + 4{,}625 \text{ ft}^3 + 1{,}100 \text{ ft}^3$$

$$\text{Volume}_{\text{Total}} = \left(7{,}600 \text{ ft}^3 \right) \left(\frac{1 \text{ yd}^3}{27 \text{ ft}^3} \right) = 281 \text{ yd}^3$$

Asphalt Pavement

Asphalt pavement consists of a **Base Course** paved with one or more layers of **Hot-Mix Asphalt** (HMA), which may be topped with a final wear course. A **Prime Coat** may be used between the base course and HMA. A **Tack Coat** is used between layers of HMA. The base course and HMA are estimated by determining the volume in cubic yards and converting it to tons. The wear course may be estimated by the ton or by the square yard. The tack and prime coats are liquids and are estimated by determining the number of gallons based on a coverage rate and converting it to tons.

Example: Determine the quantity of base, HMA, and prime coat needed to pave a 100-foot by 200-foot parking lot. The pavement requires 8 inches of base, 3 inches of HMA, and 0.15 gallons of prime coat per square yard. The density of base is 2.1 tons per cubic yard; of HMA is 2.0 tons per cubic yard; of prime coat is 8.1 pounds per gallon.

$$\text{Base} = (100 \text{ ft})(200 \text{ ft})\left(\frac{8 \text{ in}}{12 \text{ in/ft}}\right)\left(\frac{1 \text{ yd}^3}{27 \text{ ft}^3}\right)$$

$$\text{Base} = (494 \text{ yd}^3)(2.1 \text{ tons/yd}^3) = 1{,}037 \text{ tons}$$

$$\text{Prime} = (100 \text{ ft})(200 \text{ ft})(9 \text{ yd}^3/\text{ft}^3)(0.15 \text{ gal/yd}^2)$$

$$\text{Prime} = (333 \text{ gal})\left(\frac{8.1 \text{ lbs}}{1 \text{ gal}}\right)\left(\frac{1 \text{ ton}}{2{,}000 \text{ lbs}}\right) = 1.35 \text{ tons}$$

$$\text{Asphalt} = (100 \text{ ft})(200 \text{ ft})\left(\frac{3 \text{ in}}{12 \text{ in/ft}}\right)\left(\frac{1 \text{ yd}^3}{27 \text{ ft}^3}\right)$$

$$\text{Asphalt} = (185 \text{ yd}^3)(2.0 \text{ tons/yd}^3) = 370 \text{ tons}$$

Underground Utilities

Underground utilities include **Strom Drain**, **Sanitary Sewers**, **Waterlines**, and **Conduits** (for power and communication) buried in trenches. The volume of trench excavation is estimated by dividing the trench into geometric shapes or by using the **Average-End Method**. Bedding is often placed under or around the piping. The volume of bedding is calculated by calculating the cross-sectional area of the bedding and multiplying it by the length of the pipe. For large diameter pipes, the area of the pipe should be subtracted from the cross-sectional area of the bedding. The remaining trench must be backfilled with soil. The volume of the back fill is equal to the volume of the excavation less the volume of the bedding and the volume of the pipe. The volume of the pipe may be ignored for small pipes.

Example: Five hundred feet of 36-inch (38 inch outside diameter) water line is placed in a five-foot-wide trench on 12 inches of bedding. The trench is eight feet deep. Determine the required excavation volumes.

$$\text{Volume}_{\text{Excavation}} = (5 \text{ ft})(8 \text{ ft})(500 \text{ ft})\left(\frac{1 \text{ yd}^3}{27 \text{ ft}^3}\right) = 741 \text{ yd}^3$$

$$\text{Volume}_{\text{Bedding}} = (5 \text{ ft})(1 \text{ ft})(500 \text{ ft})\left(\frac{1 \text{ yd}^3}{27 \text{ ft}^3}\right) = 93 \text{ yd}^3$$

$$\text{Volume}_{\text{Pipe}} = \frac{\pi}{4}\left(\frac{38 \text{ in}}{12 \text{ in/ft}}\right)^2 (500 \text{ ft})\left(\frac{1 \text{ yd}^3}{27 \text{ ft}^3}\right) = 146 \text{ yd}^3$$

$$\text{Volume}_{\text{Backfill}} = 741 \text{ yd}^3 - 93 \text{ yd}^3 - 146 \text{ yd}^3 = 502 \text{ yd}^3$$